S0-CKI-796

TOPICS IN
THEORETICAL PHYSICS

TOPICS IN THEORETICAL PHYSICS

CHRISTOFER CRONSTROM
Editor

Proceedings of the
Liperi Summer School
in Theoretical Physics
1967

Issued originally by
Research Institute for Theoretical Physics
University of Helsinki

GORDON AND BREACH, SCIENCE PUBLISHERS
NEW YORK • LONDON • PARIS

Published 1969 by GORDON AND BREACH, Science Publishers, Inc.
150 Fifth Avenue, New York, N. Y. 10011
Library of Congress Catalog Card Number: 68-58848

Editorial office for Great Britain:

Gordon and Breach, Science Publishers Ltd.
12 Bloomsbury Way
London W.C.1

Editorial office for France:

Gordon & Breach
7-9 rue Emile Dubois
Paris 14^{e}

Distributed in Canada by:

The Ryerson Press
299 Queen Street West
Toronto 2B, Ontario

Printed in the United States of America

PREFACE

The Third Finnish Summer School in Physics, perhaps better known as the Liperi Summer School in Theoretical Physics, was held at the Liperi Vocational School in eastern Finland during the period July 27 - August 3, 1967. It was organized by a committee consisting of Professors K. V. Laurikainen (Chairman), P. Jauho, P. Lipas, P. Tarjanne, Dr. N. Mustelin and Dr. C. Cronström (Secretary), in cooperation with the Research Institute for Theoretical Physics at the University of Helsinki and the Northern Karelian Summer University, Joensuu.

The lectures at the school fell into two main categories, elementary particle physics and many-body theory. The lectures by Professor P. G. de Gennes, "Self-consistent field calculations; some topics in the theory of long molecules" and by Professor D. J. Thouless, "Excitations of many-body systems" were to be published elsewhere, and it was agreed, with regret, that publication in this volume would be redundant. All the lecture notes have been prepared by the authors themselves.

It is a pleasure to thank the president of the Liperi Vocational School, Mr. Eero Keränen, for invaluable assistance and great hospitality. Financial support from the Ministry of Education, NORDITA, the town of Joensuu and from the community of Liperi is gratefully acknowledged.

Helsinki, December 1967

Christofer Cronström
Secretary of the
Summer School

LIST OF PARTICIPANTS

H. Abarbanel Lecturer	Palmer Physical Laboratory Princeton University Princeton, New Jersey, USA
F. Abdullah	Physics Department Queen Mary College, London, England
M. B. Ali	School of Physics University of Newcastle United Kingdom
V. Ambegaokar	Laboratory of Atomic and Solid State Physics Cornell University Ithaca, New York, USA
A. Aurdal	Fysisk Institutt Avd. C Universitetet i Bergen Norway
A. Aurela	Wihurin fysiikantutkimus- laitos Turun Yliopisto Finland
T. Berggren	Inst. f. Matem. Fysik Tekniska Högskolan Lund, Sweden

A. Bialas	Instytut Fizyki Uniwersytetu Jagellonskiego Krakow, Poland
M. Blazek	Niels Bohr Institutet Copenhagen Denmark
I. Brevik	NORDITA Copenhagen Denmark
G. E. Brown Lecturer	Princeton University Princeton, New Jersey USA
L. M. Brown Lecturer	Department of Physics Northwestern University Evanston, Ill. USA
S. O. Bäckman	NORDITA Copenhagen Denmark
C. G. Callan, Jr. Lecturer	Department of Physics Harvard University Cambridge, Mass. USA
V. Celli	Department of Physics University of Virginia Charlottesville, Virginia USA
C. Cronström Secretary	Research Institute for Theoretical Physics Univ. of Helsinki Finland

U. Cronström Secr. Assistant	
B. Enflo	Inst. f. teoretisk fysik Stockholm Sweden
K. Fagerlund	Fysikaliska Institutet Abo Akademi Abo, Finland
K. G. Fogel	Fysikaliska Institutet Abo Akademi Abo, Finland
H. N. Firth	Department of Mathematics University of Durham United Kingdom
B. Giraud	Niels Bohr Institutet Copenhagen Denmark
P. G. de Gennes Lecturer	Université de Paris Faculté des Sciences - Orsay, France
S. Hameed	Theor. physics department University of Manchester England
J. Hamilton Lecturer	NORDITA Copenhagen Denmark
P. Hautojärvi	Dept. of Technical Physics Institute of Technology Otaniemi, Finland

L. Indorato	Istituto di Fisica Dell' Universita Palermo, Italy
P. Jauho	Dept. of Technical Physics Institute of Technology Otaniemi, Finland
L. Josephsen	Niels Bohr Institutet Copenhagen Denmark
J. Kurkijärvi	Université de Paris Faculté des Sciences Orsay, France
L. E. Lundberg	Inst. f. teoretisk fysik Umea Universitet Sweden
H. Miettinen	Research Institute for Theoretical Physics Univ. of Helsinki Finland
A. Molinari	Niels Bohr Institutet Copenhagen Denmark
K. Mork	Inst. for teoretisk fysikk NTH, Trondheim Norway
J. C. Parikh	Niels Bohr Institutet Copenhagen Denmark

R. Pellinen	Research Institute for Theoretical Physics Univ. of Helsinki Finland
E. Peltola	Department of Nuclear Physics Univ. of Helsinki
T. Perko	Research Institute for Theoretical Physics Univ. of Helsinki Finland
P. Pyykkö	Wihurin fysiikantutkimus-laitos, Turun Yliopisto Finland
E. Riihimäki	Department of Nuclear Physics Univ. Of Helsinki Finland
D. O. Riska	Department of Technical Physics, Inst. of Technology Otaniemi, Finland
R. Roskies Lecturer	The Weizmann Institute of Science, Department of Nuclear Physics Rehovoth, Israel
J. Savolainen	Research Institute for Theoretical Physics University of Helsinki Finland

H. Seitz	Max Planck Institut für Kernphysik - Heidelberg, Germany
P. Singer Lecturer	Department of Physics Northwestern University Evanston, Ill. USA
S. Sohlo	Department of Physics Univ. of Oulu Oulu, Finland
P. Soper	Department of Physics Surrey University London, England
E. Suhonen	NORDITA Copenhagen Denmark
P. Tarjanne	Research Institute for Theoretical Physics Univ. of Helsinki, Finland
D. J. Thouless Lecturer	Department of Mathematical Physics, University Birmingham England
N. Törnqvist	Research Institute for Theoretical Physics Univ. of Helsinki Finland
M. Vallinkoski	Research Institute for Theoretical Physics Univ. of Helsinki Finland

M. Volanen Secr. Assistant	Research Institute for Theoretical Physics Univ. of Helsinki Finland
G. Wegman	Institut für theoret. Physik, Frankfurt/Main Germany
E. Østgaard	NORDITA Copenhagen Denmark

CONTENTS

REGGE ASYMPTOTIC BEHAVIOR FOR COLLISIONS OF PARTICLES WITH SPIN

Henry Abarbanel

REGGE ASYMPTOTIC BEHAVIOR FOR COLLISIONS OF PARTICLES WITH SPIN

Henry Abarbanel
Palmer Physical Laboratory
Princeton University

I SOME REMARKS

For somewhat over a decade the complex plane has been steadily forcing its way into the lives of physicists interested in the structure of the elementary particles - especially those who turn their attention to processes involving hadrons. The primary object of interest has been the analytic structure of S-matrix elements describing scattering and decay processes. Taking as complex variables the parameters, such as scattering energy and momentum transfer, on which such S-matrix elements can depend, there has been an extensive and systematic attempt to exploit the "proven" or supposed analytic properties of such transition amplitudes to provide both insight into the properties of the particles involved and, simultaneously, concrete predictions for the values of the amplitudes themselves. The theory and phenomenology which has grown out of this massive effort is variously called "S-matrix Theory" (a likely name) or Dispersion Relations (an unlikely, but historical, name).

It must be admitted that at the time of these lectures, this theory, which is by now quite elaborate, has not exactly been a dramatic success. Although it has given us an approximate handle on some low energy scattering processes[1] and has yielded a useful set of integral representations for many of the S-matrix elements in question, its practitioners are still at a loss to extract from their equations quantitative evaluations of the basic objects under consideration.

Many of the techniques and questions involved in dispersion relations are illustrated by a somewhat historical example, that of forward scattering of light by a spinless target,[2] which we shall discuss in a moment. Besides reminding us of the usual manner in which one proceeds in this dispersing business, this example will focus our attention on the primary object of these lectures: the behavior of the S-matrix elements for large values of their arguments.

The approach we propose to adopt in answer to this question is that of Regge pole phenomelogy. By this I mean that Regge asymptotic behavior, discussed below, will be assumed to give an accurate guide to the correct asymptotic behavior of scattering amplitudes.

Following the example, we will turn our attention to the description of scattering processes involving spinning particles following the brilliant development of Jacob and Wick.[3] It is in the case of such collisions that Reggeology exhibits many of its most dramatic and possible testable consequences. Then in Section IV we will introduce the idea of Regge asymptotic behavior for transition amplitudes and make a few remarks about complex angular momentum. Subsequent to this we turn to applications of the ideas we have met and some experimental evidence on their validity. Now to the promised example.

II AN EXAMPLE

We are invited to consider the forward scattering of a photon of momentum k and polarization ϵ on a spinless target of momentum p, charge e, and mass m. The Lorentz invariant transition amplitude for the scattering from a photon helicity λ to a helicity λ' is of the form

$$T_{\lambda'\lambda}(\nu) = \vec{\epsilon}^{\,*}(\lambda') \cdot \vec{\epsilon}(\lambda)\, t(\nu)$$

where $\nu = -p.k$ is the energy variable in the problem and $T(\nu)$ is connected to the center-of-mass differential cross section at 0° by

$$\left.\frac{d\sigma}{d\Omega}(\nu, \vartheta)\right|_{\vartheta = 0} = \left|\frac{T(\nu)}{8\pi W(\nu)}\right|^2 .$$

$W(\nu)$ is the center-of-mass energy $= \sqrt{2\nu + m^2}$. The unitarity relation for $T(\nu)$ tells us that the absorptive (or imaginary) part of $t(\nu)$ has its first contribution from the lowest inelastic intermediate state. The contribution of the single particle state to $\mathrm{Im}t(\nu)$ is zero at $\vartheta = 0$ because $t(\nu)$ is even in ν. Note that we are working only to first order in α here because, among other diseases, the masslessness of the photon would plague us in interpreting the discontinuity of the scattering amplitude across its cuts as proportional to its imaginary part, which we desire. Following Gell-Mann et al.[2] we are then led to write the integral representation for $t(\nu)$, using Cauchy's theorem on the contour shown in Figure 1 and identifying

$$\mathrm{Im}\, t(\nu) = \frac{t(\nu + i\epsilon) - t(\nu - i\epsilon)}{2i} ,$$

$$t(\nu) = \frac{1}{\pi} \int_{-\infty}^{-\nu_0} d\nu' \frac{\mathrm{Im}\, t(\nu')}{\nu' - \nu}$$

$$+ \frac{1}{\pi} \int_{\nu_0}^{\infty} d\nu' \frac{\mathrm{Im}\, t(v')}{\nu' - \nu}$$

$$+ \frac{1}{2\pi i} \int_{\odot} \frac{d\nu'}{\nu' - \nu} t(\nu')$$

$\nu_0 > 0$ is the position of the inelastic contribution and the last term is the contribution from the circular contour at infinity. If $t(\nu)$ for $\nu \to \infty$ should vanish, then the final term gives zero and we have

$$t(\nu) = \frac{2}{\pi} \int_{\nu_0}^{\infty} \frac{\nu' d\nu'}{\nu'^2 - \nu^2} \mathrm{Im}\, t(\nu') ,$$

remembering that $t(\nu)$ is even.

Now for zero photon frequency $\nu \to 0$, we know that the scattering amplitude must go to the Thompson limit

$$t(0) = -2e^2 = -8\pi\alpha .$$

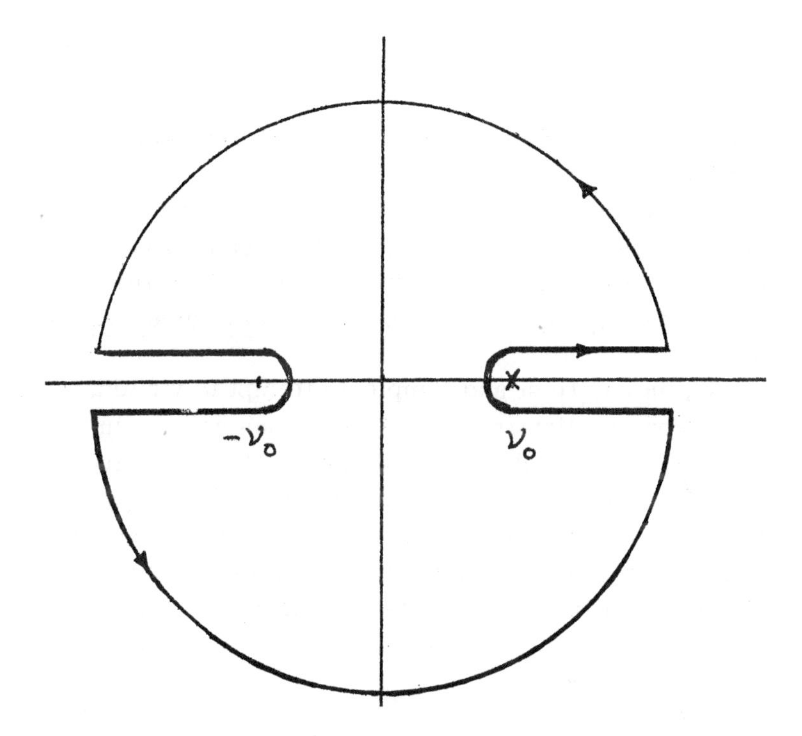

Fig. 1. The contour of integration in the ν- plane.

Further we know that the imaginary part of $t(\nu)$ is connected to the total nuclear absorption cross section[4] for photons on our spinless target by the optical theorem

$$\sigma(\nu) = \frac{\operatorname{Im} t(\nu)}{2\nu}$$

Putting these together yields

$$- \alpha = \frac{1}{2\pi^2} \int_{\nu_0}^{\infty} d\nu' \, \sigma(\nu')$$

which is a contradiction since the right hand side is positive definite.

Unless the dispersion representation for $t(\nu)$ is fundamentally false - in which case we can all go home now - the assumption above which must fail is that of its high energy (infinite ν) behavior. $t(\nu)$ cannot vanish as $\nu \to \infty$.

So, in our first and simplest attempt to write a dispersion relation free of unknown constants[5] - in this case the contribution from the circle at infinity - we have failed because of unacceptable high energy behavior. By the end of these lectures we will be able to give a Reggeistic picture of why we should have expected this failure, and where we might hope for success in the future.

III THE HELICITY FORMALISM OF JACOB AND WICK

In the past one has treated collisions of spinning particles by coupling the spin of the particles to some total spin, then coupling this total spin to the relative orbital angular momentum to form a total angular momentum J, and finally writing the S-matrix elements as sums over transitions with a definitie J. In asking questions about the polarization of individual particles, for example, this scheme is cumbersome indeed.[6] Fortunately for us Jacob and Wick recognized the enormous simplification in the scattering formalism which arises when one considers transitions between states with particles of definite spin projection along their direction of motion - the helicity.

We begin by considering the state of a particle of spin S, momentum p in the positive z direction, and helicity $\lambda = \langle S_z \rangle$:

$$| p, \lambda \rangle .$$

By applying a suitable rotation operator to this state we can describe another state with the same helicity λ and with momentum $\vec{p}$ of the same magnitude p but pointing in the direction ϑ, φ:

$$| \vec{p}, \lambda \rangle = e^{-i\varphi J_z} e^{-i\vartheta J_y} e^{-i\varphi J_z} | p, \lambda \rangle .$$

The relative phases of the various helicity states may be set for <u>massive</u> particles by Lorentz transforming $| p, \lambda \rangle$ to $p = 0$ and applying the angular momentum raising or lowering operators

$$J_{\pm} | 0, \lambda \rangle = \sqrt{(J \mp \lambda)(J \pm \lambda + 1)} \; | 0, \lambda \pm 1 \rangle .$$

For massless particles there are only two helicity states, $\lambda = \pm$ s for a particle of spin s, which can be related by a reflection in the x - z plane $Y =$ $= P e^{-i\pi J y}$

$$Y | p, \lambda \rangle = \eta | p, -\lambda \rangle .$$

η is a phase factor to be identified with the intrinsic parity of the particle.

Next let us imagine making a Lorentz transformation L on our state $| \vec{p}, \lambda >$ for massive particles, and let L take p_μ into p'_μ. L applied to our state takes it into a linear combination of states with momentum p'

$$L \mid p, \lambda > = \sum_{\lambda'} \mid L p' \lambda' X L p \lambda' \mid L \mid p \lambda >$$

$$= \sum_{\lambda'} \mid p', \lambda' >$$

$$< 0 \lambda' \mid [G^{-1}(\vec{p}') \, L \, G(\vec{p})] \mid 0 \lambda >,$$

with $G(\vec{p})$ the boosting operator which takes us from zero momentum to momentum p. $R_L = G^{-1}(L\,\vec{p})$ $L\,G(\vec{p})$ connects two states in the same frame and must be a rotation. A Lorentz transformation on $| p, \lambda >$ thus mixes up the helicities via a rotation matrix $D_{\lambda'\lambda}{}^{(s)}(R_L)$ for spin s and changes the momentum

$$L \mid p, \lambda > = \sum_{\lambda'} D^{(s)}_{\lambda'\lambda} (R_L) \mid p', \lambda' > .$$

Now we construct product states of two particles of mass m_1 and m_2 and spin s_1 and s_2. Since we will work primarily in the barycentric system where

$$\vec{p}_1 = -\vec{p}_2 = \vec{p},$$

say, we define (following the phase conventions of Jacob and Wick) the state of "particle 2" (or sometimes the "target" particle) moving in the minus z direction

$$| \overline{p, \lambda_2} > = (-1)^{s_2 - \lambda_2} e^{-i\pi J_y^{(2)}} | p, \lambda_2 > .$$

Product states for the two particles are defined by

$$| p, \lambda, \lambda_2 > = | p, \lambda, > | \overline{p, \lambda_2} > .$$

For a product state with relative momentum in the ϑ, φ direction we make an appropriate rotation generated by the <u>total angular momentum</u> of the system

$$\vec{J} = \vec{J}_1 + \vec{J}_2$$

$$| \vec{p}, \lambda, \lambda_2 > = e^{-i\varphi J_z} e^{-i\vartheta J_y} e^{i\varphi J_z} | p, \lambda_1 \lambda_2 > .$$

Jacob and Wick now show us how to construct states of definite angular momentum J and J_z by projecting from the state $| \vec{p}, \lambda_1 \lambda_2 >$

$$| J, J_z : p, \lambda_1 \lambda_2 > = \sqrt{\frac{2J+1}{4\pi}}$$

$$\int \frac{\sin\beta \, d\beta \, d\alpha \, d\gamma}{2\pi} D^{(J)*}_{J_z \lambda} (\alpha,\beta,\gamma) \, R_{\alpha\beta\gamma} \, | \vec{p}, \lambda_1 \lambda_2 >$$

where $R_{\alpha\beta\gamma}$ is a rotation through the Euler angles

$$0 < \alpha < 2\pi, \ 0 < \beta < \pi, \ 0 < \gamma < 2\pi, \ \lambda = \lambda_1 - \lambda_2,$$

and

$$D^{(J)}_{\lambda'\lambda}(\alpha, \beta, \gamma) = e^{-i\lambda'\alpha} d^{J}_{\lambda'\lambda}(\beta) e^{-i\lambda\gamma}.$$

The normalization factor in front of the integral has been chosen so

$$< J'J'_z, \lambda'_1\lambda'_2 | JJ_z\lambda_1\lambda_2 > = \delta_{J'J}\delta_{J'_z J_z}\delta_{\lambda'_1\lambda_1}\delta_{\lambda'_2\lambda_2} .$$

With this warm up we are prepared to consider the S-matrix. Our reward for the patience needed to face the many rotation matrices will soon be forthcoming We imagine that two particles of mass m_a and m_b, spin s_a and s_b, and helicities a and b scatter into two particles with labels c and d. The incoming particles we choose to be moving along the z axis and we let the scattering occur in the x-z plane. Thus $\vartheta = \varphi = 0$ for the initial product state, and $\varphi = 0$, $\vartheta =$ the scattering angle, for the final state. Our concern is for the Lorentz invariant scattering matrix

$$a_{cd,ab}(s,t)$$

which is related to the S matrix by

$$S(a + b \to c + d) = \delta_{ab,cd} + i(2\pi)^4 \delta(p_a + p_b - p_c - p_d)$$

$$N_a N_b N_c N_d \, a_{cd,ab}(s,t).$$

p_i is, of course, the four momentum of particle i, and N_i is a normalization factor which I choose to be

$$\frac{1}{\sqrt{2\,p_0}}$$

for bosons and

$$\sqrt{\frac{m_i}{p_0}}$$

for fermions. The variables s and t are those first defined by Mandelstam

$$S = -(p_a + p_b)^2 \text{ and } t = -(p_a - p_c)^2$$

and are the squares of the total energy and the four momentum transfer, respectively, in the barycentric system. a is the matrix element of the usual $\hat{T}$ operator

$$a_{cd,ab}^{(s,t)} = \langle \vartheta\ \varphi = 0\ cd \mid \hat{T}(s) \mid \vartheta = \varphi = 0\ ab \rangle$$

$$= \sum_{J,J_z} \langle \vartheta\ \varphi = 0\ cd \mid J\,J_z\ cd \rangle \langle JJ_z cd \mid \hat{T}(s) \mid JJ_z ab \rangle$$

$$\langle JJ_z\ ab \mid \vartheta = 0\ \varphi = 0\ ab \rangle .$$

From the definition of $\mid J, J_z, a\,b \rangle$ given above one quickly finds

$$< \vec{p}\ ab \mid J J_z ; p, ab > \ = \sqrt{\frac{2J+1}{4\pi}}\ D^{(J)*}_{J_z \lambda} (\varphi, \vartheta, -\varphi)$$

with $\lambda = a - b$. In our expression for $a_{cd,ab}(s,t)$ this yields

$$a_{cd,ab}(s,t) = \sum_J (2J+1)\ a^J_{cd,ab}(s)\ d^J_{\lambda\mu}(\vartheta)$$

where we have set $\mu = c - d$ and used $d^J_{\lambda,\mu}(0) = \delta_{\lambda\mu}$. This is the crucial result of Jacob and Wick.

This partial wave expansion reduces to the usual one when, say, all the particles are spinless. Then $d_{00}{}^J(\vartheta) = P_J(\vartheta)$ and

$$a_{00,00}(s,t) = \sum_J (2J+1)\ a^J_{00,00}(s,t)\ P_J(\cos\vartheta) .$$

For spinning particles the slightly more involved object $d^J_{\lambda\mu}(\vartheta)$ replaces $P_J(\cos\vartheta)$.

In real scattering processes one observes transitions between prepared states of definite quantum numbers, for example, parity. Since we will want to examine such states when we come to speak about Regge poles, I will list the properties of our product states under the discrete transformations P and T and under particle exchange. The derivations of these results are not difficult and are given by Jacob and Wick.

For the partial wave states the parity operation yields

$$P \mid JJ_z, ab > \ = \eta_a \eta_b (-)^{J - s_a - s_b} \mid JJ_z, -a, -b >$$

and time reversal

$$T \mid JJ_z, ab > = (-)^{J - J_z} \mid J - J_z, ab>$$

If we have identical particles, the exchange operator P_{ab} gives

$$P_{ab} \mid JJ_z, ab > = (-)^{J - 2s_a} \mid JJ_z, ba >$$

Using the symmetry properties of the $d^J_{\lambda\mu}(\vartheta)$

$$d^J_{\lambda\mu}(\vartheta) = d^J_{-\mu,-\lambda}(\vartheta) = (-)^{\lambda - \mu} d^J_{\mu\lambda}(\vartheta)$$

$$= (-)^{J + \lambda} d^J_{\lambda,-\mu}(\pi - \vartheta),$$

one may derive the following consequence of the invariance of the S-matrix under parity transformations

$$a_{cd,ab}(s,t) = \eta_a \eta_b \eta_c \eta_d (-)^{s_c + s_d - s_a - s_b} (-)^{\lambda - \mu}$$

$$a_{-c-d,-a,-b}(s,t),$$

and from the invariance under T,

$$a_{cd,ab}(s,t) = (-)^{\lambda - \mu} a_{ab,cd}(st).$$

Finally, let us note that the $d^J_{\lambda\mu}(\vartheta)$ can be written as

$$(\sqrt{2} \sin \frac{\vartheta}{2})^{|\lambda - \mu|} (\sqrt{2} \cos \frac{\vartheta}{2})^{|\lambda + \mu|}$$

times a polynomial in $z = \cos \vartheta$ which we will call $e^J_{\lambda\mu}(z)$.[7] In the s channel (where $a + b \to c + d$) $z_s = \cos \vartheta_s$ is a polynomial in s and t divided by a function of s alone. It is plausible then that dividing our amplitude by the given factor of

$$\sin \frac{\vartheta}{2} \text{ 's and } \cos \frac{\vartheta}{2} \text{ 's}$$

leaves us with an object free of extraneous singularities (so called kinematic singularities) in the variable t which arise because of the manner in which we have made our partial wave decomposition. Of course, the sum over $e^J_{\lambda\mu}(Z_s)$ might develop one of these extraneous singularities which was not present in any term, but as a working rule we will expect that

$$\bar{a}_{cd,ab}(s,t) = a_{cd,ab}(s,t) / \left(\sqrt{2} \sin \frac{\vartheta_s}{2}\right)^{|\lambda - \mu|} \left(\sqrt{2} \cos \frac{\vartheta_s}{2}\right)^{|\lambda + \mu|}$$

$$= \sum_J (2J + 1)\, a^J_{cd,ab}(s)\, e^J_{\lambda\mu}(Z_s)$$

is free of t kinematic singularities.[7]

To make some of this, now extensive, formalism a bit more transparent we will treat as an example pion-nucleon scattering. Since isospin is irrelevant to this discussion, we ignore it. For convenience, take the nucleon as the "projectile" and the pion as "particle 2" or the target and consider the angular momentum states

$$| J\, J_z, \lambda\, 0 \rangle$$

λ is the nucleon helicity and may be $\pm 1/2$. We can construct two states of definite parity

$$\frac{1}{\sqrt{2}}\left[\,|JJ_z, \tfrac{1}{2}\,0> \pm |JJ_z, -\tfrac{1}{2}\,0>\right],$$

with parity $\mp(-1)^{J-1/2}$ respectively. The physical partial wave transitions are thus:

$$t_1^J(s) = \frac{1}{2}\Big\{<\tfrac{1}{2}\,0\,|\hat{T}^J|\,\tfrac{1}{2}\,0> + <-\tfrac{1}{2}\,0\,|\hat{T}^J|\,-\tfrac{1}{2}\,0>$$

$$+ <\tfrac{1}{2}\,0\,|\hat{T}^J|\,-\tfrac{1}{2}\,0> \quad + <-\tfrac{1}{2}\,0\,|\hat{T}^J|\,\tfrac{1}{2}\,0>\Big\},$$

and

$$t_2^J(s) = \frac{1}{2}\Big\{<\tfrac{1}{2}\,0\,|\hat{T}^J|\,\tfrac{1}{2}\,0> + <-\tfrac{1}{2}\,0\,|\hat{T}^J|\,-\tfrac{1}{2}\,0>$$

$$- <\tfrac{1}{2}\,0\,|\hat{T}^J|\,-\tfrac{1}{2}\,0> - <-\tfrac{1}{2}\,0\,|\hat{T}^J|\,\tfrac{1}{2}\,0>\Big\}.$$

Using time reversal invariance these reduce to

$$t_{1,2}^{\,J}(s) = \left[<-\tfrac{1}{2}\,0\,|\hat{T}^J|\,\tfrac{1}{2}\,0> \pm <\tfrac{1}{2}\,0\,|\hat{T}^J|\,\tfrac{1}{2}\,0>\right].$$

Now consider the two independent (using P and T again) helicity amplitudes

$$a_{\frac{1}{2}0,\frac{1}{2}0}(s,t) = \sum_J{}' \,(2J+1)<\tfrac{1}{2}\,0\,|\hat{T}^J|\tfrac{1}{2}\,0>\, d^J_{\frac{1}{2},\frac{1}{2}}(\vartheta_s)$$

and

$$a_{-\frac{1}{2}0,+\frac{1}{2}0}(s,t) = \sum_J (2J+1) \langle -\tfrac{1}{2}\, 0 \,|\hat{T}^J|\, +\tfrac{1}{2}\, 0 \rangle\, d^J_{-\frac{1}{2},+\frac{1}{2}}(\vartheta_s)$$

These are sums over $[t_1^J(s) \pm t_2^J(s)]/2$ and represent certain combinations of physical transitions. In experiments one would measure the a's and learn about resonances, for example, by studying the $t_i^J(S)$'s.

The connection between the helicity formalism and the usual one is found by considering,[8]

$$a_{\lambda' 0,\lambda 0}(s,t) = \bar{u}(p',\lambda') \left[-A(s,t) + i\,\gamma\left(\frac{k'+k}{2}\right) B(s,t)\right] u(p,\lambda)$$

where A and B are the C G L N scalar invariants, k and k' are the incident and final pion momenta and $u(p,\lambda)$ is a helicity spinor for a nucleon of momentum p, helicity λ. When the nucleon is the projectile, an explicit representation for the spinors is

$$u(p,\lambda) = \sqrt{\frac{p_0+m}{2m}} \left[\frac{2|p|\lambda}{p_0+m}\right] \chi_\lambda$$

where

$$\sigma \cdot \hat{p}\, \chi_\lambda = 2\lambda \chi_\lambda$$

and

$$\bar{u}(p',\lambda') = \sqrt{\frac{p_0' + m}{2m}}\ \chi^+_{\lambda'}\, e^{i\frac{\vartheta_s}{2}\sigma_y}\left(1, -\frac{2p'\lambda}{p_0' + m}\right)$$

The rotation matrix

$$e^{i\frac{\vartheta_s}{2}\sigma_y}$$

reminds us that the outgoing nucleon is moving in the direction $(\sin\vartheta_s, 0, \cos\vartheta_s)$. We know that $a_{\lambda'0,\lambda 0}(s,t)$ may also be written in the form

$$a_{\lambda'0,\lambda0}(s,\ t) =$$

$$[\chi^+_{\lambda'}\, e^{i\frac{\vartheta_s}{2}\sigma_y}\,(f_1(s,\ t) + \sigma\cdot\hat{p}'\,\sigma\cdot\hat{p}\, f_2(s,\ t))\ \chi_\lambda]$$

$$= [f_1(s,\ t) + 4\lambda'\lambda\, f_2(s,\ t)]\ [\ |\lambda' + \lambda|\cos\frac{\vartheta_s}{2} + (\lambda' - \lambda)\sin\frac{\vartheta_s}{2}\ ]$$

or

$$a_{\frac{1}{2}0,\frac{1}{2}0}(s,\ t) = \cos\frac{\vartheta_s}{2}\ (f_1 + f_2)$$

and

$$a_{-\frac{1}{2}0,\frac{1}{2}0}(s,\ t) = -\sin\frac{\vartheta_s}{2}\,(f_1 - f_2)$$

The connection of $a_{\lambda'0,\lambda0}(s,t)$ is now given when we express f_1 and f_2 in terms of A and B

$$f_1(s,t) = \frac{p_0 + m}{2m} \left[A + (\sqrt{s} - m)B \right]$$

and

$$f_2(s,t) = \frac{p_0 - m}{2m} \left[-A + (\sqrt{s} + m)B \right]$$

Note that it is

$$a_{1/20,1/20} / \cos\frac{\vartheta_s}{2} \quad \text{and} \quad a_{-1/20,1/20} / \sin\frac{\vartheta_s}{2}$$

which are proportional to A(s,t) and B(s,t) the scalar amplitudes free of kinematic singularities

$$\frac{a_{1/20,1/20}}{\cos\frac{\vartheta_s}{2}} = A + \frac{s - m^2 - m_\pi^2}{2m} B$$

and

$$\frac{a_{-1/20,1/20}}{\sin\frac{\vartheta_s}{2}} = \frac{1}{2m\sqrt{s}} \Big[A\,(s + m^2 - m_\pi^2) + m\,B\,(s + m_\pi^2 - m^2) \Big]$$

but that only the t kinematic singularities have been removed by this procedure.

Finally, from parity considerations one may deduce the connection between the $t_i^J(s)$ given before and the conventional transitions

$$f_{l\pm}(s) = f_{J\pm 1/2}(s)$$

$$t_1^J(s) = f_{J-1/2}(s)$$

$$t_2^J(s) = f_{J+1/2}(s) .$$

IV REGGEOLOGY LORE

The basic premise of Reggeology is that a partial wave amplitude $a^J(s)$ representing some physical transition may be given for non-integers (or half integer) J by a unique interpolating function $a(J,s)$ which equals $a^J(s)$ for physical values of J and is meromorphic (has only poles) in the J plane. The positions of the poles are, in general, functions of s, say $\alpha(s)$, although one is now bold enough to consider the possibility of poles which do not move with s. Some years ago Regge[9] was able to prove these properties in non-relativistic potential scattering theory. It almost goes without saying that in spite of extensive and heroic effort no one has succeeded in doing the same for relativistic scattering processes. In fact, at the very least we know that due to the inelasticity of relativistic scattering there could (and probably do) appear moving cuts in J. In the following we will consider only poles.

One of the primary attractions[10] of such moving

poles is that they provide a connection among particles with different masses and spins but the same internal quantum numbers, for as s varies, $\alpha(s)$ occasionally passes through an integer or half-integer, α_0. At those points $a(J,s)$ has a pole for J equal to α_0, and since $a(J,s)$ is equal to $a^J(s)$ for such values of J, the physical transition amplitude has a pole, representing some particle. As $\alpha(s)$ varies, therefore, it sweeps out a trajectory on which some particles lie. One can go further yet and imagine that not only do all the observed particles lie on Regge trajectories but also that the only trajectories around are those on which the observed particles lie. We should, as we use this idea, remember to recognize it as a strong assumption.

Let us now look at some of the physical consequences of these statements. To begin we consider the scattering of spinless equal mass m particles for which the transition amplitude is

$$a(s,t) = \sum_J (2J+1)\, a^J(s)\, P_J(z_s)$$

in the s-channel or

$$= \sum_J (2J+1)\, b^J(t)\, P_J(z_t)$$

in the t-channel center of mass. The second series, which we will now concentrate on, converges in a certain ellipse in z_t which includes the t physical region

$$t \geq 4m^2, \quad |z_t| \leq 1.$$

In terms of s and t

$$z_t = \frac{s-u}{t-4m^2} = \frac{2s+t-4m^2}{t-4m^2} = \frac{\nu}{q_t^2} .$$

As the next step we examine the partial wave amplitude in a little more detail by inverting the series for a(s,t)

$$b^J(t) = \frac{1}{2} \int_{-1}^{+1} dz_t \, P_J(z_t) \, a(s,t).$$

For large J, $P_J(Z_t)$ is not sufficiently well behaved to guarantee the uniqueness of an interpolating function so this representation will not do for our b(J,t). If we write a fixed t dispersion relation for a(s,t),[12]

$$a(s,t) = \frac{1}{\pi} \int_{s_0}^{\infty} \frac{ds'}{s'-s} a_s(s',t) + \frac{1}{\pi} \int_{u_0}^{\infty} \frac{du'}{u'-u} a_u(u',t)$$

where $a_i(s,t)$ is the absorptive part of a(s,t) in the ith channel, we find

$$b^J(t) = \frac{1}{\pi} \int_{x_0}^{\infty} \frac{ds'}{2q_t^2} Q_J(z_t(s)) \{ a_s(s',t) + (-)^J a_u(s',t) \}$$

with $x_0 = \min(S_0, U_0)$. We used the conventional definition of the Legendre function of the second kind

$$Q_J(z) = \frac{1}{2} \int_{-1}^{+1} \frac{dx}{z - x} P_J(x)$$

The large J behavior of the $Q_J(z)$ is quite sufficient for our interpolation

$$Q_J(z) \to \sqrt{\frac{\pi}{2J}} \frac{e^{-(J+\frac{1}{2})\phi}}{(\sinh \phi)^{1/2}} \quad \text{as } |J| \to \infty$$

for $|\arg (Z + 1)| \leq \pi$, $|\arg J| < \pi$ and $\phi = \cosh^{-1} Z$, so that if we define, following Froissart and Gribov,[11]

$$b^{\pm}(J,t) = \frac{1}{\pi} \int_{x_0}^{\infty} \frac{ds}{2q_t^2} Q_J(z_t) \, [\, a_s(s',t) \pm a_u(s',t) \,],$$

we rid ourselves of the additional unpleasant large J behavior of $(-1)^J$ and have a function suitable for interpolating between integer J's. It is the $b^{\pm}(J,t)$ that we will assume to contain the Regge poles; they are usually called the signatured amplitudes. For even (odd) J the signatured amplitude $b^{\pm}(J,t)$ coincides with the physical transition.

Now we are prepared to consider the partial wave series again. We rewrite the sum as a contour integral around the real axis as in Figure 2

$$a(s,t) = \frac{i}{2} \int_C dJ\,(2J+1)\,\frac{1}{\sin \pi J}\, b^J(t)\, P_J(-z_t)$$

$$= \frac{i}{2} \int_C \frac{dJ(2J+1)}{\sin \pi J} \left[\frac{1+e^{-i\pi J}}{2}\, b^+(J,t) + \frac{1-e^{-i\pi J}}{2}\, b^-(Jt) \right] P_J(-z_t)$$

This trick is called a Sommerfeld-Watson transformation and can clearly be used to convert any sum over a discrete index into a contour integral. Regge poles enter the picture when we open up the contour and push it back into the left half plane. Assuming we get nothing from the large circle at infinity (true in potential scattering) and letting only one Regge pole enter each of $b^{\pm}(J,t)$ in the form

$$b^{\pm}(J,t) = \frac{\beta^{\pm}(J,t)}{J - \alpha_{\pm}(t)} + \text{regular part},$$

there results

$$a(s,t) = \pi\, \frac{(2\alpha_{+}(t) + 1}{\sin \pi\, \alpha_{+}(t)}\, \beta^{+}(\alpha_{+}(t),t)\, \frac{1+e^{-i\pi\alpha_{+}(t)}}{2}\, P_{\alpha_{+}(t)}(-z_t)$$

$$+\pi \frac{(2\alpha_-(t)+1)}{\sin\pi\alpha_-(t)} \beta^-(\alpha_-(t),t) \frac{1-e^{-i\pi\alpha_-(t)}}{2}$$

$$P_{\alpha_-(t)}(-z_t)$$

$$+\frac{i}{2}\int_{L-i\infty}^{L+i\infty} dJ \frac{2J+1}{\sin\pi J} b(J,t) P_J(-z_t) \quad \text{with } L < 0$$

Now for large values of s, which takes us out of the t physical region where the original partial wave series converged we are invited to consider large z_t for which

$$P_J(z_t) \sim \frac{2^J \Gamma(J+\frac{1}{2})}{\sqrt{\pi}\,\Gamma(J+1)} z_t^J + o(z_t^{J-2})$$

Thus if $\alpha_+(t)$ or $\alpha_-(t)$ is greater than L, we may neglect the integral and have at large s

$$a(s,t) \sim \bar{\beta}^+(\alpha_+(t),t) \frac{1+e^{-i\pi\alpha_+(t)}}{\sin\pi\alpha_+(t)} s^{\alpha_+(t)}$$

$$+\bar{\beta}^-(\alpha_-(t),t) \frac{1-e^{-i\pi\alpha_-(t)}}{\sin\pi\alpha_-(t)} s^{\alpha_-(t)}$$

and for the absorptive part in s

$$a_s(s,t) \sim \bar{\beta}^+ s^{\alpha_+(t)} + \bar{\beta}^- s^{\alpha_-(t)}$$

This is, if we further allow ourselves to continue t to physical values in the s channel, $t \leq 0$, the typical form for Regge asymptotic behavior for large s and fixed t.

The standard conjecture for the asymptotic behavior of a scattering amplitude can now be cast in the form: for fixed momentum transfer and large energy the amplitude is given by the Regge form above where α (t) refers to possible trajectories which could give rise to physical states with the quantum numbers of the t(or crossed) channel. One must examine the allowed contributions and choose those with the largest value of $\alpha(t)$ to be the dominant contributions.

It is precisely in this manner that we shall use the Regge formalism, namely as a guide to the asymptotic behavior of scattering amplitudes. Other features of the Regge formula are bound up in the knowledge of the Regge residue functions which are not likely to be given a satisfactory theoretical treatment for some time.

In the case of spinning particles there is one additional feature that we shall enjoy exploiting. As we have learned, the replacement for $P_J(z)$ in the partial wave expansion is $d^J_{\lambda\mu}(\vartheta)$. When we enter the realm of complex angular momentum, we will occasionally encounter situations in which $\alpha(t)$ takes on integral values less than max (λ,μ). Such a point corresponds to a total angular momentum less than its projection on some axis. Since such a thing is completely unphysical, it is given the affectionate name of a "nonsense" point.[7] Thus, for example, when we encounter an object like $d_{20}(t)$ we expect to find zeroes in the helicity amplitude corresponding to it at $\alpha(t) = 1, 0, \ldots$.

Let's now return to π-N scattering and see how this formalism goes through there. To examine the large s behavior of the two s-channel helicity amplitude $a_{\pm,+}(s,t)$ (drop the 0 helicity of the pions and denote $\pm 1/2$ by $\pm$), we are instructed to have a look into the t-channel reaction $N\overline{N} \to \pi\pi$. Of course, there are also two amplitudes here which we will call $A_{+\pm}(s,t)$. The connection between the a's and the A's can be written as[13]

$$a_{+,+}(s,t) = \cos\chi(s,t)\, A_{+-}(s,t) + \sin\chi(s,t)\, A_{++}(s,t)$$

and

$$a_{-,+}(s,t) = -\sin\chi(s,t)\, A_{+-}(s,t) + \cos\chi(s,t)\, A_{++}(s,t)$$

and

$$\cos\chi(s,t) = \frac{(s + m_N^2 - m_\pi^2)\sqrt{t}}{\sqrt{[s-(m_N+m_\pi)^2][s-(m_N-m_\pi)^2]}\,\sqrt{t-4m_N^2}}$$

The partial wave helicity states for the $N\overline{N}$ system are four in number (suppressing the J and J_z):

$$|++>_\pm = \frac{1}{\sqrt{2}}\,(|++> \pm |-->),\ P = \pm(-)^J,\ C = (-)^J$$

and

$$|+->_\pm = \frac{1}{\sqrt{2}}\,(|+-> \pm |-+>),\ P = \pm(-)^J, C = \pm(-)^J.$$

The charge conjugation and parity quantum numbers to the right of the states tell us that $|++>_-$ and $|+->_-$ are respectively the $^1(J)_J$ and $^3(J)_J$ $N\bar{N}$ states, while $|+\pm>_+$ are mixtures of the $^3(J\pm 1)_J$ states. The parity of the π-π state is $(-1)^J$ giving us two physical transitions

$$m_1^J(t) \equiv \frac{\sqrt{2}}{2} < 00 \mid \hat{T}^J(t) \mid ++ >_+$$

$$= < 00 \mid \hat{T}^J \mid ++ >$$

and

$$m_2^J(t) \equiv \frac{\sqrt{2}}{2} < 00 \mid \hat{T}^J(t) \mid +- >_+$$

$$= < 00 \mid \hat{T}^J \mid +- >$$

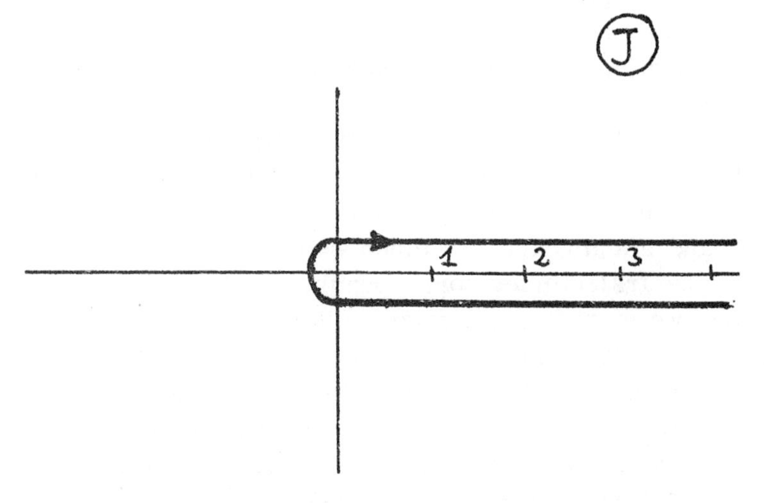

Fig. 2. The contour in the J-plane for the Watson-Sommerfeld Transformation.

Thus $A_{+\pm}(s.t)$ are already sums over physical partial wave amplitudes:

$$A_{++}(s,t) = \sum_J (2J+1)\, m_1^J(t)\, d_{00}^J(\vartheta_t) = \sum_J (2J+1)\, m_1^J(t)\, P_J(z_t)$$

$$A_{+-}(s,t) = \sum_J (2J+1)\, m_2^J(t)\, d_{10}^J(\vartheta_t)$$

For isospin 0 in the t-channel we have only even J, and for isospin 1, only odd.

We can already see one feature of the π-N Regge formula by looking at the partial wave expansion of A_{+-}. Since it contains a $d^J{}_{10}(\vartheta_t)$ the contribution of any given Regge trajectory will vanish when $\alpha(t) = 0$, and if at some s value one Regge pole dominates the amplitude, then the full amplitude should become quite small (of background integral size) at $\alpha(t) = 0$. Thus for fixed large s where a one Regge pole formula is applicable, we should expect a dip in the differential cross section at t varies through the value where $\alpha(t) = 0$.

Concentrate now on the I = 1 configuration in the t channel which corresponds to the $a^{(-)}$ amplitudes in the s channel. The G parity must be + 1, J must be odd, and P and C are -1. Since we have agreed to seek out those trajectories corresponding to physical particles, we must find a particle with these quantum numbers. The only candidate on my wallet card is the ρ meson. So only one trajectory contributes here, namely $\alpha_\rho(t)$. We may, in fact, now determine the trajectory function by looking at the differential cross section for πN charge exchange, $\pi^- p \to \pi^0 n$, which is proportional to the $a^{(-)}$ transition. Each of $a^{(-)}_{\pm,+}(s,t)$ behaves for large s as

$$\gamma(t) s^{\alpha_\rho(t)} \frac{(1 - e^{-i\pi\alpha_\rho(t)})}{\sin \pi \alpha_\rho (t)}$$

which, as far as the exponential behavior goes, gives for the differential cross section

$$\frac{d\sigma(\pi^- p \to \pi^0 n)}{dt} = \tilde{\gamma}(t)\, s^{2(\alpha_\rho(t) - 1)}$$

from which $\alpha_\rho(t)$ may be extracted. The experiments[14] show that $\alpha_\rho(t)$ is a straight line over the measured region $-1(\text{BeV})^2 \leq t \leq 0$ and that extrapolating along a straight line one arrives at $\alpha_\rho = 1$ near $t = m_\rho^2$, where, of course, α_ρ should be one.

$$\frac{d\alpha_\rho(t)}{dt} \approx \frac{1}{(\text{BeV})^2}$$

Further, and of special interest to us here, $\alpha_\rho(t) = 0$ near $t \approx -0.6(\text{BeV})^2$, and at that point

$$\frac{d\sigma}{dt}$$

for πN charge exchange exhibits a significant dip. In the absence of any other arguments for the presence of the dip, this constitutes a strong indication that something quite Reggelike is going on here.[15]

Another consequence of this trajectory is a prediction for the rate of decrease of the difference in the total cross section for $\pi^+ p$ and $\pi^- p$ elastic scattering. $\sigma(\pi \pm p) - \sigma(\pi^- p)$ is proportional to Im $a^{(\mp)}_{+}, + (s,0)$ and thus behaves for large s as

$$\sigma(\pi^+ p) - \sigma(\pi^- p) \sim \text{constant } s^{\alpha_\rho(0) - 1}$$

$$\sim \text{constant } s^{-0.5}$$

which seems to agree with the observed manner in which these two cross sections approach one another.

In order to incorporate into the phenomenology the observed fact that the total cross sections for $\pi^\pm p$ $K^\pm p$, pp, and other processes tend to constants for large s, one has introduced a trajectory which has the quantum numbers of the vacuum and is endowed with the wonderous property that α (0) = 1. This very handy construct is called the Pomeranchuk trajectory, $\alpha_p(t)$. It, sad to tell, seems to have no particles (yet found) which lie upon it, and it is often told that this is because this very special trajectory has a slope at most 1/2 or 1/3 that of all other trajectories with particles on them, which seem to have

$$\frac{d\alpha(t)}{dt} \sim \frac{1}{(\text{BeV})^2}.$$

As our original concern was with learning about the possibility of writing unsubtracted dispersion relations, we may now examine this for πN scattering in the light of the lore just presented. The two amplitudes

$$\bar{A}_{++}(s,t) = A_{++}(s,t)$$

and

$\overline{A}_{+-}(s,t) = A_{+-}(s,t)/\sin\vartheta_t$ are free of s kinematic singularities, and are thus acceptable for fixed t dispersion relations. For I = 0 in the t channel the asymptotic behaviors are for large s

$$\overline{A}^{(0)}_{++}(s,t) \sim s^{\alpha_P(t)} \text{ and } \overline{A}^{(0)}_{+-}(s,t) \sim s^{\alpha_P(t)-1}$$

Since $\overline{A}^{(0)}_{+-}(s,t)$ is odd under s,u crossing, or under the operation

$$\nu = \frac{s - u}{4} \rightarrow -\nu ,$$

it will satisfy an unsubtracted dispersion relation for fixed $t \leq 0$, $\overline{A}^{(0)}_{++}(\nu,t)$ is even and needs one subtraction. For I = 1 in the t channel $\overline{A}^{(1)}_{+\pm}(\nu,t)$ is odd (even) and behaves as $\nu^{\alpha_\rho(t)}$ $(\nu^{\alpha_\rho(t)-1})$ for large ν (or s). Thus each of these amplitudes satisfies an unsubtracted fixed t dispersion relation for (at least) $t \leq 0$. One subtraction, therefore, suffices to give a complete set of fixed t dispersion relations for πN scattering, for once we have determined the t-channel helicity amplitudes, we may evaluate the s-channel a's using the crossing relation given above.

AN APPLICATION

Having discussed pion-nucleon scattering we have exhausted the most interesting example as far as physics goes. Our next example will thus serve to illustrate a number of points. We examine once again the Compton scattering of photons on spinless targets (called pions of mass m whose helicities we do not exhibit. There are two independent helicity amplitudes in the s channel: $a_{1,1}(s,t)$ and $a_{1,-1}(s,t)$. In the forward direction only $a_{1,1}(s,0)$ survives and is then equal to $T(\nu)$ of the example in Section II. The crossing relations for pion Compton scattering are extremely simple and up to an overall phase read

$$a_{1,1}(s,t) = A_{1-1}(s,t)$$

$$a_{1,-1}(s,t) = A_{11}(s,t)$$

where A's, as usual, refer to the t channel, $\gamma\gamma \to \pi\pi$ The partial wave decomposition of $A_{1-1}(s,t)$ is

$$A_{1-1}(s,t) = \sum_{J=2}^{\infty} (2J+1)\, a^{J}_{1-1}(t)\,(p_t^2 k_t^2 \sin^2 \vartheta_t)$$

$$P''_J(Z_t),$$

and

$$p_t = \sqrt{\frac{t}{4} - m^2}\,, \quad k_t = \sqrt{\frac{t}{4}}$$

and are respectively the pion and photon momenta in the t-channel center of mass.

For pions of isospin one we may have $I_t = 0$, 1, or 2 in the t channel $I_t = 0$ and 2 correspond to J even (or even signature in the Regge language), and $I_t = 1$ must have J odd (or odd signature) because of Bose statistics. Of the trajectories which can contribute to

$A_{1-1}(s,t)$ the leading ones have I = 0 and are the Pomeranchuk, and the one on which the $f^0(1250)$ meson lies, call it $\alpha_f(t)$. The predicted asymptotic behavior for $a_{1,1}(s,t)$ is now

$$a_{1,1}(s,t) \underset{s \to \infty}{\to} \beta_p(t)\alpha_p(t)(\alpha_p(t)-1)s^{\alpha_p(t)} \frac{(1+e^{-i\pi\alpha_p(t)})}{\sin \pi \alpha_p(t)}$$

$$+ \beta_f(t)\alpha_f(t)(\alpha_f(t)-1)s^{\alpha_f(t)} \frac{(1+e^{-i\pi\alpha_f(t)})}{\sin \pi \alpha_f(t)}$$

Since $a_{1,1}(s,t)$ is even in

$$\nu = \frac{s - u}{4}$$

for $I_t = 0$ and 2, it will not satisfy an unsubtracted fixed t dispersion relation until $\alpha_p(t) < 0$ which, if it occurs, is quite far from the range of physical interest, $t \lesssim 0$. At $t = 0$, $\alpha_p(0) = 1$, and unless $\beta_p(t)$ is singular as $1/t$, the contribution of the vacuum trajectory vanishes (another example of a nonsense zero) and the leading behavior is

$$s^{\alpha_f(0)} \sim s^{0.4}$$

which still is not sufficient for ''no-subtraction''. This behavior certainly confirms the results of Section II where we showed that $t(\nu)$ could not satisfy an unsubtracted dispersion relation.

It seems appropriate now to inquire whether there is any amplitude for the pion Compton effect for which an unsubtracted fixed t dispersion relation holds. An examination of $a_{1,-1}(s,t)$ shows that it is not an acceptable candidate. Let us, however, turn to the lore we have about removing kinematic singularities of amplitudes. By our rules, dividing $a_{1,1}(s,t)$ by $1 + \cos \vartheta_s$ removes the t kinematic singularities from it. Similarly, dividing $A_{1-1}(s,t)$ by $\sin^2\vartheta_t$ removes the s kinematic singularities from it. The crossing relation between

$$\frac{a_{1,1}}{1 + \cos \vartheta_s}$$

and $A_{1-1}/\sin^2\vartheta_t$ is

$$\frac{a_{1,1}(s,t)}{1 + \cos \vartheta_s} = \frac{\sin^2\vartheta_t}{1 + \cos \vartheta_s} \frac{A_{1-1}(s,t)}{\sin^2\vartheta_t}$$

$$= - \frac{(s - m^2)^2}{8p_t^2 k_t^2} \frac{A_{1-1}(s,t)}{\sin^2\vartheta_t}$$

so that

$$\frac{a_{1,1}(s,t)}{\left(\frac{s - m^2}{2}\right) (1 + \cos \vartheta_s)} = \frac{a_{1,1}(s,t)}{st + (s - m^2)^2}$$

is free of both s <u>and</u> t kinematic singularities.[16] This new amplitude, call it $\hat{a}(s,t)$, has better asymptotic behavior than $a_{1,1}(s,t)$ and by our Regge arguments should satisfy an unsubtracted fixed t dispersion relation

$$\hat{a}(s,t) = \frac{2e^2}{(m^2 - s)(m^2 - u)} + \frac{1}{\pi} \int_{s_0}^{\infty} ds' \, \hat{a}_s(s',t) \left[\frac{1}{s' - s} + \frac{1}{s' - u} \right],$$

using the fact that $\hat{a}$ is even in ν. This yields for $a_{1,1}(\nu,0)$

$$a_{1,1}(\nu) = -2e^2 + \frac{2\nu^2}{\pi} \int_{\nu_0}^{\infty} \frac{d\nu' \, \mathrm{Im}\, a_{1,1}(\nu')}{\nu'(\nu'^2 - \nu^2)}$$

$$= -2e^2 + \frac{4\nu^2}{\pi} \int_{\nu_0}^{\infty} \frac{d\nu'}{\nu'^2 - \nu^2} \sigma(\nu')$$

which is perfectly consistent with the Thompson limit $a_{1,1}(0) = -2e^2$ and is just the good old Kramers-Kronig dispersion relation.[2]

Before leaving this example let us dwell on the consequence of the vanishing of the Pomeranchuk contribution to $a_{1,1}(s,0)$. Since

$$\sigma(s) = \frac{1}{(s - m^2)} \operatorname{Im} a_{1,1}(s,0)$$

the total nuclear cross section[4] will vanish as $s^{\alpha_f(0)-1}$ unless $\beta_p(t)$ becomes singular or unless the partial wave amplitude has a <u>fixed</u> pole at J = 1 which would provide the 1/J - 1 just needed to cancel the nonsense zero. It is not only unpleasant but also inconsistent with a simple version of inelastic unitarity which says $\operatorname{Im} a_{1,1}(s,0)$ does not fall off at large s faster than s/log s, for $\sigma(s)$ to vanish as a power of s for large s. It behooves us, then, to examine the partial wave amplitude $\hat{a}(J,t)$ for $I_t = 0$ in the expansion

$$\hat{A}(s,t) = \frac{A_{1-1}(s,t)}{st + (s - m^2)^2} = \sum_J (2J + 1)\, \hat{a}(J,t)\, P_J''(Z_t).$$

The neighborhood of J = 1 will interest us especially. Inverting the series gives

$$\hat{a}(J,t) = \frac{1}{(2J + 1)(2J - 1)(2J + 3)} \frac{1}{2} \int_{-1}^{1} dz_t\, \hat{A}(z_t,t)$$

$$[(2J + 3) P_{J-2}(z_t) - 2(2J + 1) P_J(z_t) + (2J - 1) P_{J+1}(z_t)]$$

$z_t = \nu/p_t k_t$, so writing a fixed t dispersion relation in

ν for $\hat{A}(\nu,t) = \hat{a}(\nu,t)$ yields for the signatured $\hat{a}^{(+)}$ (J,t) up to an overall irrelevant constant,

$$\hat{a}^{(+)}(J,t) = \int_{\nu_0}^{\infty} d\nu \, \mathrm{Im}\, \hat{a}(\nu,t)$$

$$[(2J+3)Q_{J-2}(z_t) - 2(2J+1)Q_J(z_t) + (2J-1) Q_{J+2}(z_t)]$$

The $Q_e(Z)$ have poles when $e = -1,-2 \ldots$ so near $J = 1$, $a^{(+)}(1,t)$ has a fixed pole with residue proportional to

$$R(t) = \frac{1}{\pi} \int_{\nu_0(t)}^{\infty} d\nu \, \mathrm{Im}\, \hat{a}(v,t) = \frac{2e^2}{t} \left(\frac{2}{\sqrt{3}} \right)$$

$$+ \frac{1}{\pi} \int_{\nu_i(t)}^{\infty} d\nu' \, \mathrm{Im}\, \hat{a}(r',t),$$

where the one particle contribution has been explicitly separated out ($2/\sqrt{3}$ is a Wigner coefficient) and $\nu_i(t)$ is the inelastic threshold.

If $R(t)$ vanishes, one can show $\beta_p(t)$ must be singular and $\sigma(\nu)$ must approach

$$\frac{4\pi e^2}{3} \frac{d}{dt} \alpha_p(t) \Big/_{t=0}$$

for large ν.[17] If R(t) does not vanish, we have a fixed pole which restores the Pomeranchuk contribution and at the same time yields constant total cross sections. This type of result is not restricted to spinless targets. On targets of isospin I, hypercharge Y, if R(t) vanishes, the total asymptotic cross section is

$$\sigma(\nu)_{\nu\to\infty} = 2\pi e^2 \alpha'_p(0) \left[\frac{Y^2}{4} + \frac{I(I+1)}{3} \right].$$

Finally, let us look at the configuration with $I_t = 2$. $\hat{a}$ (s,t) for this isospin behaves at large s as

$$s^{\alpha_{I_4 = 2}(t) - 2}.$$

Since there are no observed I = 2 particles, it has been suggested[18] that $\alpha_{I=2}(0) < 0$ so that

$$a_{1,1}^{I_t = 2}(s,0)$$

should vanish as $s \to \infty$. By constructing the Kramers-Kronig relation for $I_t = 2$ and requiring that

$$a_{1,1}^{I_t = 2}(s,0)$$

vanish for large s one derives the following "super-convergence" relation

$$-2e^2 = \frac{4}{\pi} \int_{\nu_0}^{\infty} d\nu' \left[\sigma^{\gamma\pi^+}(\nu') - \sigma^{\gamma\pi^0}(\nu')\right],$$

which is the same as a sum rule found by Harari and Pagels.[19] In the presence of a fixed pole at J = 0, which is quite likely there,[17] this sum rule or superconvergence relation cannot hold.

Homework:

1. Examine the scattering of vector meson on scalar mesons. Enumerate the helicity amplitudes, exhibit their crossing relations, and evaluate their high energy behavior. Let the scattering particles all have isospin one, and write down any superconvergence relations that might follow from

$$\alpha_p(0) < 1 \text{ and } \alpha_{I=2}(0) < 0.$$

2. Work out the Reggeology for Compton scattering on a proton. Examine the t-channel partial wave amplitudes in the neighborhood of J = 1 and derive the sum rules which follow from the vanishing of the residual of fixed poles there. Call one of these the Drell-Hearn[20] sum rule. What amplitudes for this process satisfy unsubtracted dispersion relations.

VI SOME MORE REMARKS

This concludes our introduction to Regge folklore. I have purposely passed over interesting problems connected with unequal mass scattering and daughter trajectories, for example,[21] and concentrated on the phenomenological use of Regge asymptotic behavior to inform us about the possibility of unsubtracted dispersion relations. This seems the best established consequence of the Regge apparatus and, as we have now seen, is sufficient to lead us to a number of physically interesting results. It is well to retain a healthy amount of skepticism about even this aspect of Reggeology since it is possible that the whole game may vanish one day. In case it does not, the presentation here may be of some value. In the meantime, it remains our only systematic and reliable guide to the asymptotic behavior of S-matrix elements.

REFERENCES AND FOOTNOTES

1. G. F. Chew and F. E. Low, Phys. Rev. 107, 1570 (1956).
2. M. Gell-Mann, M. L. Goldberger, and W. Thirring, Phys. Rev. 95, 1612 (1954).
3. M. Jacob and G. C. Wick, Ann. Phys. (N. Y.) 7, 404 (1959).
4. Pair production must be ignored to lowest order in α.
5. An "unsubtracted" dispersion relation.
6. In the book by M. L. Goldberger and K. M. Watson, Collision Theory, J. Wiley and Sons, New York (1964), the "ancient" L-S coupling scheme is presented in some detail. Contrast the treatment of nucleon-nucleon scattering given these with the development of M. L. Goldberger, M. T. Grisaru, S. W. MacDowell, and D. Y. Wong, Phys. Rev. 120, 2250 (1960).
7. M. Gell-Mann, M. L. Goldberger, F. E. Low, E. Marx, and F. Zachariasen, Phys. Rev. 133, B 145 (1964).

8. G. F. Chew, M. L. Goldberger, F. E. Low, and Y. Nambu, Phys. Rev. 106, 1337 (1957).
9. T. Regge, Nuovo Cimento 14, 951 (1959) and ibid 18, 147 (1960).
10. G. F. Chew and S. C. Frautschi, Phys. Rev. Letters 7, 394 (1961) and ibid 8, 41 (1962).
11. The semi-mathematical literature on complex angular momentum is large. Most of the basic results and a bibliography is given by R. Oehme, Strong Interactions and High Energy Physics, Scottish Universities' Summer School 1963, Plenum Press, New York (1964) and by E. J. Squires, Complex Angular Momentum and Particle Physics, W. A. Benjamin, Inc., New York (1963).
12. To do this a(s,t) should be free of s kinematic singularities. This is accomplished for helicity amplitudes by the method described above.
13. T. L. Trueman and G. C. Wick, Annals of Physics (N. Y.) 26, 322 (1964).
14. A good summary of experiments and theory on Regge dips due to nonsense transitions is given by S. C. Frautschi, "New Experimental Evidence on Regge Poles in High Energy Scattering," a lecture given at the Second Tokyo Summer Institute of Physics.
15. Honesty forces us to admit that the "exchange" of a single Regge pole, as here, leads also to the prediction that there is no polarization of the outgoing nucleon. Unfortunately, the outgoing nucleons have a measured non-vanishing polarization at the energies under consideration. Perhaps this is due to the interference of the Regge term with some residual resonant background or

possibly is caused by cuts. It could, of course, also indicate that the whole phenomenology we have constructed is incorrect.

16. D. Horn, private communication. L. L. Wang, Phys. Rev. 142, 1187 (1966).
17. H. Abarbanel, F. E. Low, I. J. Muzinich, S. Nussinov, and J. H. Schwarz, to appear in Phys. Rev., and A. H. Mueller and T. L. Trueman, to appear in Phys. Rev.
18. V. de Alfaro, S. Fubini, G. Furlan, and G. Rosetti, Phys. Lett. 21, 576 (1966), and F. J. Gilman and H. Harari, Phys. Rev. Letters 18, 1150 (1967).
19. H. Pagels, Phys. Rev. Letters 18, 316 (1967) and H. Harari, ibid, 319.
20. S. D. Drell and A. C. Hearn, Phys. Rev. Letters 16, 908 (1966).
21. M. L. Goldberger and C. E. Jones, Phys. Rev. 150, 1269 (1966), and D. Z. Freedman and J. M. Wang, Phys. Rev. Letters 17, 569 (1966).

MANY-BODY THEORY

G. E. Brown

MANY BODY THEORY

G. E. Brown
Princeton University

I. INTRODUCTION

The graphical techniques introduced by Feynman into field theory prove to be very useful in many-body theory. These techniques were introduced by Goldstone (Proc. Roy. Soc. A 239, 267 (1957)) and Hugenholtz (Physica 23, 481 (1957)). Goldstone uses an expansion of the development operator,

$$U(t) = e^{-iHt} \qquad \text{(we set } \hbar = 1) \qquad \text{I(1)}$$

whereas Hugenholtz expands the resolvent

$$R(z) = \frac{1}{H - z}. \qquad \text{I(1.1)}$$

These two operators are related by

$$R(z) = -i \int_{-\infty}^{0} U(t) e^{izt}\, dt \text{ for } \mathrm{Im}(z) < 0,$$
$$R(z) = i \int_{0}^{\infty} U(t) e^{izt} dt \text{ for } \mathrm{Im}(z) > 0. \qquad \text{I(1.2)}$$

The expansion of $U(t)$ is easier to use, since we generally have to deal with products of functions rather than with the convolutions of functions which arise in the expansion of $R(z)$. The resolvent has a more direct physical interpretation, since the poles of its matrix elements give the eigenvalues of the Hamiltonian.

We again split the Hamiltonian into

$$H = H_0 + H_1 \qquad \text{I(2)}$$

and then expand in powers of H_1.

The use of the development operator is connected with the interaction representation, and I think it relevant to discuss the latter, although one could make the expansion of (1) without ever introducing it.

Consider the time-dependent Schrödinger equation

$$\frac{d\Psi}{dt} = -i(H_0 + H_1)\Psi$$

$$\Psi(t) = e^{-iHt}\Psi(0) = U(t)\Psi(0). \qquad \text{I(3)}$$

Introduce the interaction representation by

$$\hat{\Psi}(t) = e^{iH_0t}\Psi(t) = e^{iH_0t}U(t)\hat{\Psi}(0)$$

$$= \hat{U}(t)\hat{\Psi}(0) \qquad \text{I(3.1)}$$

with the development operator

$$\hat{U}(t) = e^{iH_0 t}\, U(t). \qquad I(3.2)$$

We have used

$$\Psi(0) = \hat{\Psi}(0). \qquad I(3.3)$$

More generally, the development operator carrying $\hat{\Psi}$ from t' to t is

$$\hat{U}(t,t') = e^{iH_0 t}\, U(t - t')\, e^{-iH_0 t'}. \qquad I(3.4)$$

Then

$$\frac{d\hat{\Psi}}{dt} = -ie^{iH_0 t} H_1 e^{-iH_0 t}\, \hat{\Psi} = -iH_1(t)\hat{\Psi} \qquad I(3.5)$$

with

$$H_1(t) \equiv e^{iH_0 t} H_1 e^{-iH_0 t}.$$

In going over to the interaction representation, we are simply making a unitary transformation. We have reached a wave function $\hat{\Psi}$, whose rate of change depends only on the perturbing part of the Hamiltonian. The operator $\hat{U}(t)$ is just as convenient to use in calculating transition amplitudes as $U(t)$, since its matrix elements differ only trivially, i.e.,

$$(\Phi_n, \hat{U}(t)\, \Phi_m) = e^{iE_n t}\, (\Phi_n, U(t)\Phi_m). \qquad I(4)$$

(See Dirac, Quantum Mechanics, 3rd and 4th ed., where in § 44 a, U(t), called T*there, is discussed in detail).

We know from (3.1) that

$$\frac{d}{dt}\hat{U}(t) = -iH_1(t)\,\hat{U}(t) \qquad \text{I(5)}$$

with

$$\hat{U}(0) = U(0) = 1$$

Thus,

$$\hat{U}(t) = 1 - i\int_0^t H_1(t_1)\,dt_1 + (-i)^2 \int_0^t dt_2 \int_0^{t_2} dt_1 H_1(t_2) H(t_1) + \ldots \qquad \text{I(6)}$$

The reason that we must use this relatively complicated form of expansion of the exponents in $\hat{U}(t)$ is that $H_1(t)$ does not commute with $H_1(t')$ for $t' \neq t$, and we must, therefore, keep careful track of the order of factors. We can easily check that (6) is a solution of (5) by taking d/dt of (6). Alternatively, we can express

$$\hat{U}(t) = \sum_{n=0}^{\infty} \frac{(-i)^n}{n!} \int_0^t \cdots \int_0^t T\,[H_1(t_n)\cdots H_1(t_2)H_1(t_1)]\, dt_1 dt_2 \cdots dt_n \qquad \text{I(7)}$$

and, more generally,

$$\hat{U}(t,t') = \sum_{n=0}^{\infty} \frac{(-i)^n}{n!} \int_{t'}^{t} \cdots \int_{t'}^{t} T[H_1(t_n) \cdots H_1(t_2) H_1(t_1)]$$

$$dt_1 dt_2 \cdots dt_n \qquad \text{I(7.1)}$$

The letter T followed by a bracket denotes a product of the operators inside the bracket rearranged so that the arguments decrease from left to right. This is called a time-ordered product.

We shall apply this perturbation theory to calculate either properties of the ground state or of states which do not differ much from the ground state. It is convenient to treat the state Φ_0 as the normal state of affairs and talk only about particles added to it or particles missing from it (holes). In this way we treat it as the unperturbed, or bare, vacuum in field theory.

We shall use the letter i to denote states with momentum less than the Fermi momentum k_F, and the letter m to denote states with momentum greater than k_F, and k when we don't specify whether the state is that of a particle or hole. We rewrite the creation and annihilation operators a_k^+ and a_k as

$$a_m = a_m, \quad a_m^+ = a_m^+ \quad \text{For } m > k_F$$

$$\text{I(8)}$$

$$a_i = b_i^+, \quad a_i^+ = b_i \quad \text{For } i < k_F$$

and call b_i^+ the creation operator for a hole and b_i the annihilation operator. We now have

$$a_m \Phi_0 = b_i \Phi_0 = 0 \qquad \text{I(8.1)}$$

$$\text{all } m > k_F \text{ and } i < k_F$$

so that Φ_0 is the state with no particles and no holes in it; eqs. (8.1) bring out the analogy with the bare vacuum in field theory clearly.

We now take the unperturbed Hamiltonian to be

$$H_0 = \sum_k T_k a_k^+ a_k = \sum_{i < k_F} T_i b_i b_i^+$$

$$+ \sum_{m > k_F} T_m a_m^+ a_m \qquad \text{I(9)}$$

and the perturbation to be

$$H_1 = \frac{1}{2} \sum V_{k_1 k_2, k_3 k_4} a_{k_2}^+ a_{k_1}^+ a_{k_3} a_{k_4} . \qquad \text{I(9.1)}$$

We introduce now a graphology for the matrix elements of $\hat{U}(t)$. By way of example, let us look at the second-order matrix element of $\hat{U}(t)$ in (6). This is given by*

$$(\Phi_0, H_1(t_2)\,\Phi_n)(\Phi_n, H_1(t_1)\Phi_0) \qquad \text{I(10)}$$

with

$$\Phi_n = a^+_{m_2} a^+_{m_1} b^+_{i_2} b^+_{i_1} \Phi_0. \qquad \text{I(10.1)}$$

(For the moment we will not worry about the order of the a's and b's. Later on we shall introduce conventions for this.) We draw time as going upward, a particle as a line going upwards and a hole as a line directed downwards. Thus, Φ_n can be represented as

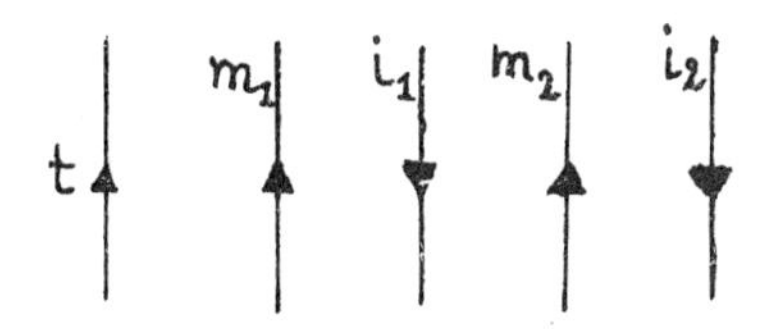

The action of $H_1(t_1)$ on Φ_0 in (10) must be then such as to create particles in m_1 and m_2 and holes in i_1 and i_2. Then $H_1(t_2)$ must annihilate the same particles and holes, so that (10) can be represented in the way illustrated in Fig. 1, where the dashed line represents the potential V. The matrix element of V represented by the lower dashed line in the direct term in Fig. 1a is

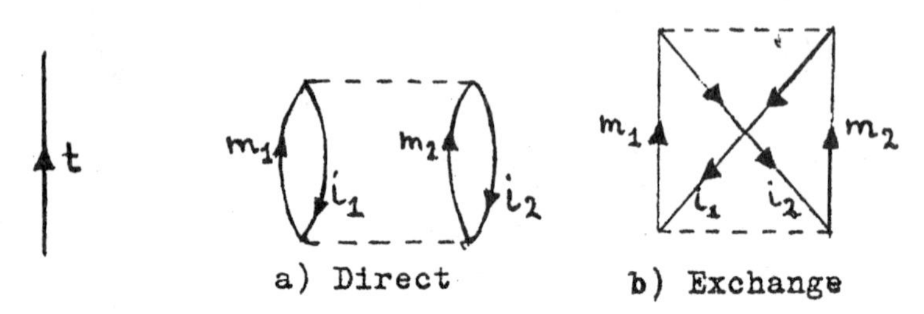

Fig. 1. Graphical illustration of the second-order matrix elements.

The ↑ represents a particle; ↓ a hole and - - - - - - the potential.

$$(m_1 m_2 \,|V|\, i_1 i_2) \qquad \text{I}(10.2)$$

and that represented by the upper dashed line is

$$(i_1 i_2 \,|V|\, m_1 m_2), \qquad \text{I}(10.3)$$

or the complex conjugate of (10.2). In the Exchange term, Fig. 1b, the lower dashed line represents (10.2) but the upper dashed line represents

$$(i_2 i_1 \,|V|\, m_1 m_2)$$

Fig. 1a gives the contribution*

$$(-i)^2 \,|(m_1 m_2 \,|V|\, i_1 i_2)|^2 \, e^{-i(T_{m_1} + T_{m_2} - T_{i_1} - T_{i_2})(t_2 - t_1)} \qquad \text{I}(11)$$

*We shall develop rules for determining numerical factors later.

whereas Fig. 1b gives

$$-(-i)^2 (i_2 i_1 \mid V \mid m_1 m_2)\, e^{-i(T_{m_1}+T_{m_2}-T_{i_1}-T_{i_2})(t_2-t_1)}$$

$$(m_1 m_2 \mid V \mid i_1 i_2) \qquad \text{I}(11.1)$$

where the (-) sign enters automatically into the exchange term because of the commutation relations of the a's.

In order to see how this relates back to the ordinary time-independent theory, one should carry out the integrations over time indicated in (6). This is done in Section 46 of the 3rd and 4th edition of Dirac's Quantum Mechanics.

In defining $H_1(t)$ by (3.2), we have implicitly given our a's and b's a time dependence. This is seen by inserting a factor

$$e^{-iH_0 t}\, e^{+iH_0 t} = 1$$

between each operator in H_1, so that

$$H_1(t) = \frac{1}{2} \sum V_{k_1 k_2, k_3 k_4}\, a^+_{k_2}(t)\, a^+_{k_1}(t)\, a_{k_3}(t)\, a_{k_4}(t) \qquad \text{I}(11.2)$$

with

$$a_k(t) \equiv e^{+iH_0 t}\, a_k\, e^{-iH_0 t} \qquad \text{I}(11.3)$$

Expanding out the exponentials, one finds

$$a_k(t) = a_k + i\ [H_0, a_k]\ t + i^2 \frac{1}{2} [H_0, [H_0, a_k]] t^2 + \cdots$$

$$= a_k - iT_k t a_k + \frac{1}{2} i^2 T_k^2 t^2 a_k + \cdots \qquad \text{I(11.4)}$$

$$= a_k e^{-iT_k T}$$

where T_k is the number defined in (9). An easy way of remembering the minus sign in (11.4) is to note that a_k operating on a wave function diminishes the number of particles, so that the operation of the H_0 in the exponent on the left gives a result smaller than that of the H_0 on the right.

The use of the matrix elements (4) in the calculation of transition probabilities is quite straightforward, and is explained in Dirac's book. The calculation of energies is a bit more subtle, but we shall show that the evaluation of

$$(\Phi_0, \hat{U}(t)\ \Phi_0) = e^{iE_0 t} (\Phi_0, U(t)\Phi_0) \qquad \text{I(12)}$$

enables one to find the ground-state energy.

D. Wick's Theorem; Rules for Graphs

The analysis of the series (7) in terms of graphs is facilitated by using a theorem due to Wick, Phys. Rev. 80, 268 (1950). This theorem relates the time-ordered product of operators to the so-called "normal product".

The normal product of a number of operators a, a^+, b etc. is defined as the product of the operators rearranged so that the creation operators stand to the left of the annihilation operators, with a minus sign (coming from the Fermi character of the operators) if the rearrangement involves an odd number of permutations of the operators. For example

$$N[b_{i_1} b^+_{i_2} a_{m_1} a^+_{m_2}] = b^+_{i_2} a^+_{m_2} a_{m_1} b_{i_1}$$

$$N[b_{i_1} b^+_{i_2} a^+_{m_1} a^+_{m_2}] = - b^+_{i_2} a^+_{m_1} a^+_{m_2} b_{i_1} \qquad \text{I(13)}$$

The order of the creation operators or of the annihilation operators among themselves does not matter, since they anticommute. The distributive law

$$N[(A + B)C] = N[AC] + N[BC] \qquad \text{I(13.1)}$$

makes it possible to calculate a normal product of operators which are not just simple annihilation and creation operators, but which may be sums of them.

In our perturbation series, the interaction H_1 involves a product of four creation or annihilation operators which all relate to the same time. We define the time-ordering of these to be such that the operators a^+ (b_2 or a^+_m in our notation) stand to the left of the operators a (b^+_2 or a_m).

We denote the difference between the time-ordered product and the normal product of two operators the "contraction" of the operators, and we denote it by joining them with a line e.g.

$$T\,[UV] = \overline{UV} + N[UV]. \qquad I(14)$$

We assume that U and V are simple operators, like a, a^+. An example is

$$T[a_{m_2}(t_2)a^+_{m_1}(t_1)] = \overline{a_{m_2}(t_2)a^+_{m_1}}(t_1) + N[a_{m_2}(t_2)a^+_{m_1}(t_1)] \qquad I(14.1)$$

where

$$N[a_{m_2}(t_2)a^+_{m_1}(t_1)] = -\,a^+_{m_1}(t_1)a_{m_2}(t_2)$$

and

$$\overline{a_{m_2}(t_2)a^+_{m_1}}(t_1) = a_{m_2}(t_2)a^+_{m_1}(t_1) + a^+_{m_1}(t_1)a_{m_2}(t_2) \text{ if } t_2 > t_1 \qquad I(14.2)$$

$$\overline{a_{m_2}(t_2)a^+_{m_1}}(t_1) = 0 \text{ if } t_2 < t_1.$$

The contraction is clearly just a number, not depending on the a's and b's, since it is either an anticommutator or zero. For example, for $t_2 > t_1$, using (11.4) one gets

$$\overline{a_{m_2}(t_2)a^+_{m_1}}(t_1) = \delta_{m_1m_2} e^{-iT_{m_2}(t_2-t_1)} \qquad I(14.3)$$

If $t_2 < t_1$ this contraction is zero.

Wick's theorem states that

$$T[UV \cdots YZ] = N[UV \cdots XYZ] + N[\overline{UV} \cdots XYZ] + N[U\overline{V \cdots X}YZ] + N[\overline{U\overline{V \cdots X}Y}Z] + \cdots, \qquad I(15)$$

i.e., the time-ordered product is equal to the sum of normal orders after all possible contractions have been made. Again, in the above it is to be understood that one has a minus sign in front of the relevant term if an odd number of permutations are required to bring the two operators together in any contraction.

The usefulness of Wick's theorem can be seen by remembering that the expectation value of the normal product of any product of operators between vacuum states Φ_0 is zero, i.e.

$$(\Phi_0, N[UV]\,\Phi_0) = 0. \qquad I(15.1)$$

So that all that remains of $T[UV \cdots YZ]$ when taken between vacuum states is the sum of terms in which all operators are paired.

The proof of Wick's theorem is easily made by induction. It is obviously true if there are only two operators - then it is just the definition of the contraction. If we assume that the theorem is always true for a product of n operators, we can prove it for a product of $n + 1$ operators. Suppose that the operator with the earliest time coordinate is W, so that

$$T[UVWXYZ] = T[UVXYZ]W. \tag{I(16)}$$

If W is an annihilation operator, the product of it and any other operator is already in normal form, so that its contraction with any other operator is zero. Therefore, we can simply multiply (15) on the right and include W in the normal products, since it is standing to the right as it should. Since all contractions involving W give zero, the theorem for $n + 1$ operators is satisfied if W is an annihilation operator. If W is a creation operator, it can be commuted through to the left of all the normal products in (15) and then put inside the normal product signs. The anticommutator of W with any operator is just the contraction, so that the theorem again holds. If W is a sum of creation and annihiliation operators, the distributive law, which holds for time-ordered products, normal products and contractions, ensures that the theorem holds for any product of $n + 1$ operators. Since the theorem holds for $n = 2$, it must hold for all higher n.

Let us summarize the possible normal orders and contractions now.

$$N[a_{m_2}(t_2)a^+_{m_1}(t_1)] = -a^+_{m_1}(t_1)a_{m_2}(t_2),$$

$$\text{I(17)}$$

$$\overbrace{a_{m_2}(t_2)a^+_{m_1}}(t_1) = \delta_{m_1 m_2} e^{-iT_{m_1}(t_2-t_1)}, \quad t_2 > t_1$$

$$= 0 \qquad\qquad , \quad t_2 \leq t_1.$$

Remembering that

$$a_{i_1}(t_2)\, a^+_{i_2}(t_1) = b^+_{i_1}(t_2) b_{i_2}(t_1)$$

where $i < k_F$ and that we consider b^+ as creation and b as annihilation operator

$$N[b^+_{i_2}(t_2) b_{i_1}(t_1)] = b^+_{i_2}(t_2) b_{i_1}(t_1)$$

$$\begin{aligned} \overline{b^+_{i_2}(t_2) b_{i_1}}(t_1) &= 0 & ,t_2 > t_1 \\ &= - b_{i_1}(t_1) b^+_{i_2}(t_2) - b^+_{i_2}(t_2) b_{i_1}(t_1) \\ &= - \delta_{i_1 i_2}\, e^{-iT_{i_1}(t_2 - t_1)}, & t_2 \le t_1 \end{aligned} \qquad \text{I(18)}$$

$$\overline{a^+_{m_2} a}^{\,+}_{m_1} = \overline{a_{m_2} a_{m_1}} = \overline{a^+_m b_i} = \overline{a^+_m b}^{\,+}_i = \overline{b^+_{i_1} b}^{\,+}_{i_2}$$

$$= \overline{b_{i_1} b_{i_2}} = 0.$$

(The equal part of the $\le$ signs comes from our definition for the time-ordering of two operators referring to the same time.) These contractions may also be called "unperturbed propagators," the term "propagator" coming from Feynman's work in the theory of the electron. In a sense, the contraction

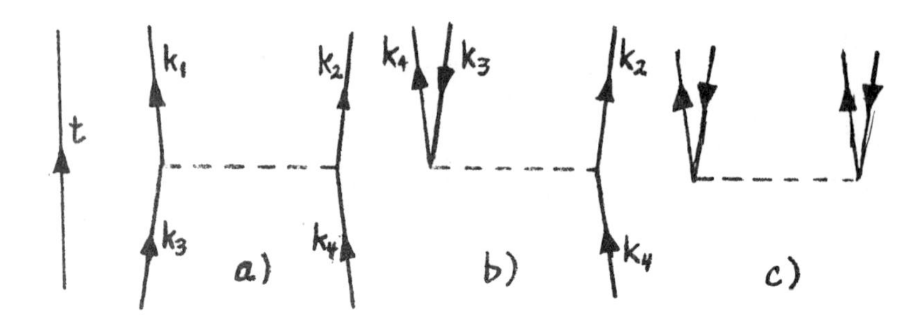

Fig. 2. Representations of

$$H_1 = \frac{1}{2}(k_1k_2|V|k_3k_4)\, a^+_{k_2} a^+_{k_1} a_{k_3} a_{k_4} .$$

describes the propagation of the particle from time t_1 to t_2, or the propagation of the hole from t_2 to t_1.

Wick's theorem makes it easy to calculate expectation values of operators in the configuration Φ_0. The expectation value of a normal product of operators is zero, since annihilation operators give zero when operating on the "vacuum" Φ_0. The only terms on the right-hand-side of (15) which survive are those in which all the operators are contracted.

The process of contraction can be represented graphically. Each H_1 in (16) is represented by such graphs as shown in Fig. 2. There a^+ is represented as a line going <u>out</u> from the vertex, and a as a line going in. Thus, in Fig. 2a, particles in states k_3 and k_4 scatter into states k_1 and k_2. In Fig. 2b, a particle in state k_4 scatters into state k_2, while a particle-hole pair, with particle in state k_1 and hole in state k_3, is created. Remember that a line going backwards in time represents a hole. The difference between 2a and 2b is clearly that in 2a, $k_3 > k_F$, so that a_{k_3} represents a particle annihilation operator, whereas in 2b, $k_3 < k_F$, so that a_{k_3} represents a hole creation operator.

Let us now think about representing the matrix element

$$(\Phi_0 | H_1(t_n) ------ H_1(t_2) H_1(t_3) | \Phi_0)$$
$$t_n > t_{n-1} > ------- t_2 > t_1, \quad \text{I(19)}$$

where, as we remarked, only the terms on the right-hand-side of (15) survive in which <u>all</u> operators are contracted. First of all let us understand the meaning of the contraction (14.3) of two operators

$$a_{m_2}(t_2) \text{ and } a^+_{m_1}(t_1).$$

This contraction is zero unless $m_2 = m_1$ and $t_2 > t_1$. It clearly means that a particle created in state m_1 at time t_1 is annihilated at time t_2. Thus, the contraction

$$a^+_{j_2} a^+_{j_1} a_{j_3} a_{j_4} (t_2) a^+_{k_2} a^+_{k_1} a_{k_3} a_{k_4} (t_1) \quad \text{I(19.1)}$$

is represented by the diagram in Fig. 3, where the equalities $j_3 = k_1$, etc., are brought about by the $\delta_{j_3 k_1}$, etc., in (17) and (18). One sums over all eight lower indices in (19.1). Thus, the contraction of $H_1(t_2)$ $H_1(t_1)$ produces factors like

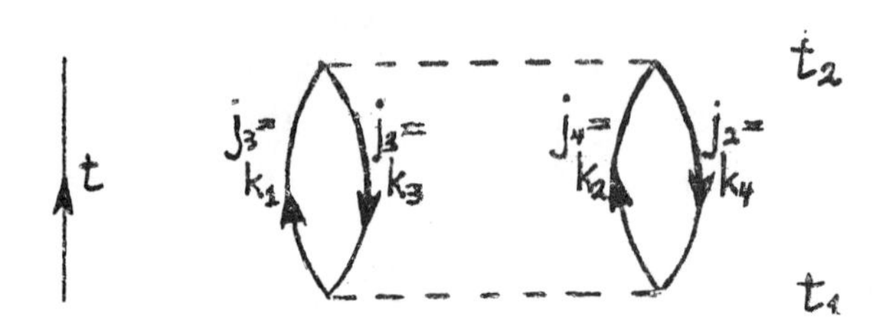

Fig. 3. Representation of the contraction in (19.1).

$$\frac{1}{4}(k_3 k_4 \,|V|\, k_1 k_2)(k_1 k_2 \,|V|\, k_3 k_4), \qquad \text{I}(19.2)$$

that is, just the same matrix elements of the potential as (10.2) and (10.3). (We have used a different notation in (19.2) from that in (10.2) and (10.3), but this signifies nothing.) There are clearly four possible pairings of (19.1). The one which gives

$$j_3 = k_2,\ j_1 = k_4,\ j_4 = k_1,\ j_2 = k_3$$

gives the same result as (19.2) so that for the total, one has twice (19.2). The two other pairings reproduce the exchange term, Fig. 1b.

In drawing the graphs it is helpful to remember that since

$$a^+_{j_2} a^+_{j_1} a_{j_3} a_{j_4}$$

is multiplied by

$$\frac{1}{2} \int \psi^*_{j_1}(x)\, \psi^*_{j_2}(x')\, V(x,x')\, \psi_{j_3}(x)\, \psi_{j_4}(x')\, d^3x d^3x'.$$

The indices j_1 and j_3 must always refer to the same point on the graph, as must j_2 and j_4. In drawing the dashed lines to represent the potentials, one is separating the points x and x' in the integrand, so that one could think of the points in the graph as being x, t_2, x',t_2 etc., although one integrates over x and x', as one will also do later over t_2 and t_1.

It can now be seen that all processes coming from the product (19) can be obtained by drawing the dashed potential lines and joining them up in all possible ways. Thus, for example, typical third-order terms are:

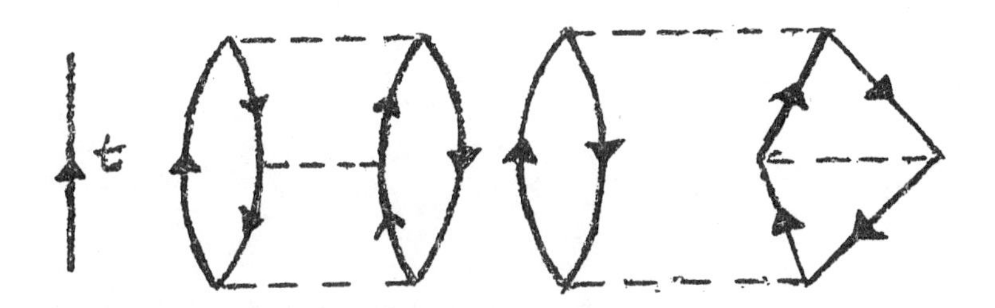

and there are very many more possible graphs one can draw, already in this order.

We note, finally, that if a line joins a point to itself, it must be a hole line, since the contraction of two particle operators vanishes if the times are equal (See (17)). A graph of particular interest is

representing the pairing

$$a^{+}_{k_2} a^{+}_{k_1} a_{k_3} a_{k_4}$$

which gives the matrix element

$$\frac{1}{2}(k_1 k_2 \, | V | \, k_1 k_2)$$

i.e., the diagonal element entering into the Hartree theory. One has, also, the exchange term

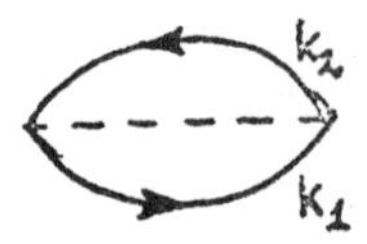

giving

$$\frac{1}{2}(k_1 k_2 \, | V | \, k_2 k_1).$$

Of course, it is not so easy to indicate whether a line joining a point with itself goes up or down; we will just remember that they always represent holes.

The perturbation series for

$$(\Phi_0, \hat{U}(t) \, \Phi_0) \qquad \text{I(20)}$$

is given in this way as a sum over graphs. The contribution of a particular graph is $-\frac{1}{2}$ i times the matrix element from each dashed potential line, the -i coming from the expansion of the exponential as in

(6). Furthermore, a factor -1 comes from each closed loop of fermion lines. These factors arise when the contractions are removed from a product; they come because an odd permutation of operators must be made.

This is a tricky point, and we discuss it in more detail. Consider a product such as (19.1). The pairs of operators referring to each given point are

$$a^+_{j_1} a_{j_3} \,, \quad a^+_{j_2} a_{j_4} \,, \quad a^+_{k_1} a_{k_3} \,, \quad a^+_{k_2} a_{k_4} \,.$$

In thinking of forming the contractions for Fig. 3, we bring together the operators first as

$$a^+_{j_1} a_{j_3} a^+_{k_1} a_{k_3} \quad ; \quad a^+_{j_2} a_{j_4} a^+_{k_2} a_{k_4}$$

where the first four operators form the left-hand loop. So far there has been no change in sign, since transferring a pair of operators from one place to another always involves an even number of permutations. In our contractions, we bring the creation operator to the right of the annihilation operator, thus, the pairing of the first four operators is

$$-\overbrace{a_{j_3} \quad a^+_{k_1}} \qquad \overbrace{a_{k_3} \quad a^+_{j_1}}$$

and it is clear that one gets a minus sign for each closed loop of particle lines.

In addition, one gets a minus sign from each contraction representing a hole propagator (See (18)).

For each time-difference between two dashed potential lines at times t_r and t_{r-1}, one has the factor

$$e^{-i(t_r - t_{r-1})D_r}, \quad t_r > t_{r-1}$$

where D_r is equal to the sum of the kinetic energies of the particle lines which cross a horizontal section of the graph between t_r and t_{r-1}, minus the sum of the kinetic energies of the hole lines crossing the same section. These factors come from insertion of the particle and hole contractions, or "propagators", (17) and (18). Thus in Fig. 3, one has the factor

$$e^{-i(T_{k_1} + T_{k_2} - T_{k_3} - T_{k_4})(t_2 - t_1)}.$$

Integrating this over time

$$\int_0^t dt_2 \int_0^{t_2} dt_1 \, e^{-i(T_{k_1} + T_{k_2} - T_{k_3} - T_{k_4})(t_2 - t_1)}$$

the upper limit on the integral over t_1 is t_2 since we have chosen a definite time ordering. This integral is easy to carry out; its value is

$$\frac{i}{(T_{k_3} + T_{k_4} - T_{k_1} - T_{k_2})} \left\{ t - \frac{e^{-i(T_{k_1} + T_{k_2} - T_{k_3} - T_{k_4})t} - 1}{i(T_{k_3} + T_{k_4} - T_{k_1} - T_{k_2})} \right\} \quad \text{I(21)}$$

As $t \to \infty$, the second term in $\{\}$'s becomes negligible compared with the first and we need to keep only

$$\frac{it}{(T_{k_3} + T_{k_4} - T_{k_1} - T_{k_2})}$$

In general, if we do perturbation theory to n^{th} order, we have n - 1 energy denominators, one coming from each integral over a relative time, and n-1 factors i in the numerator which remove all but one of the i's mentioned in the discussion following (20).

We have a factor $1/n!$ in the expansion of (7). However, there are n! permutations of the various times $t_n \cdots t_2, t_1$ among themselves, all producing identical results, which remove this factor.

Finally, there will be one time integration left over, since up to now we have integrated only over the time differences, showing that (20) is proportional to -i t, where we have kept here the one -i in the expansion of (6) which is not cancelled by the n-1 i's produced in the integrations over time differences. This argument applies only to a graph which is connected, i.e., cannot be split into two parts unconnected by particle or hole lines, since otherwise each unconnected part will be proportional to -i t. We shall discuss these unconnected parts ("unlinked graphs") in the next section.

In summary, the nth order coefficient of - i t in (29) is composed, for each nth order graph, of the factors:

1) -1 for each closed loop of particle lines

2) -1 for each hole line

3) $(\Sigma T_i - \Sigma T_m)^{-1}$ for each of the n-1 horizontal sections, where the ΣT_m is carried out over all particle lines cut, whereas ΣT_i is carried out over all hole lines

4) $\frac{1}{2}(k_1k_2 \,|V|\, k_3k_4)$ for each dashed line, where k_3 and k_1 are the lines entering and leaving the one end of the dashed line, and k_4 and k_2 are those entering and leaving the other end.

Furthermore, topologically - identical graphs may be given by various ways of pairing, and one must multiply by the number of these (see the factor 2 discussed after eq. (19.2)). The factor coming from these and the factors of 1/2 entering into 4) can be lumped simply into a common factor, both being taken care of by the revised rule. This rule was explained to me by Baird Brandow.

4') First choose a graph, preferably a "direct" one. (See Fig. 1, where both the direct and exchange second-order graphs are shown). Then insert for each dotted line the direct minus exchange matrix element

$$(k_1k_2 \,|V|\, k_3k_4) - (k_2k_1 \,|V|\, k_3k_4).$$

This gives one all "exchange-equivalent" graphs;

e.g., insertion of the above for both lines in Fig. 1a produces, also, the contribution, Fig. 1b. The rule for obtaining the common factor is, then:

Include a factor of 1/2 for each equivalent pair of lines (particles or holes). Two lines from an equivalent pair if they:

a) both begin at the same interaction,
b) both end at the same interaction, and
c) both go in the same direction.

To see how this comes about, start with the direct graph of Fig. 1

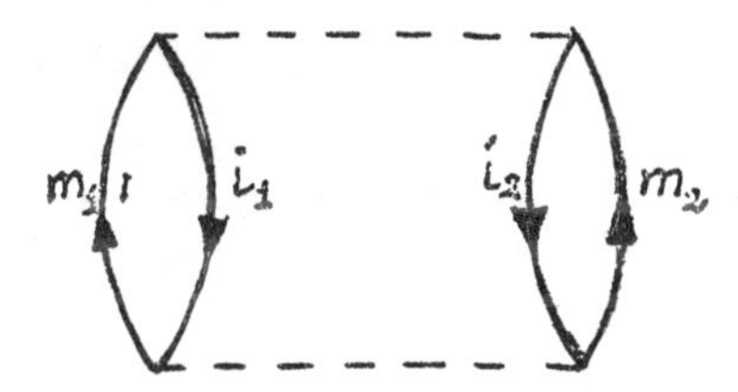

for which the contribution to the energy is

$$W^{(2)} = \frac{1}{2} \sum_{\substack{m_1, m_2 > k_F \\ i_1, i_2 < k_F}} \frac{|(m_1 m_2 | V | i_1 i_2)|^2}{T_{i_1} + T_{i_2} - T_{m_1} - T_{m_2}} . \qquad \text{I(22)}$$

The net factor of 1/2 in front of the direct and exchange term essentially comes about because in old-fashioned perturbation theory one would sum only over $m < n$ in order to produce only distinct intermediate states. Since the particles are identical (and the wave function antisymmetrical), one would double count by summing over both $m < n$ and $m > n$.

Suppose, now, one makes an insertion, such as shown below.

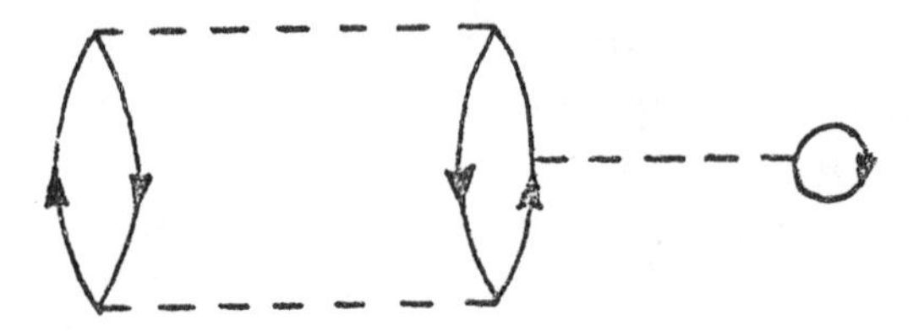

This insertion can be made into either the line labelled by m_1 or m_2 in the preceeding figure, each giving the same contribution. The net numerical factor in the above graph will consequently be 1/2. (No additional factor of 1/2 comes in the insertion, since either the $\Phi^*(x_1)\,\Phi(x_1)$ or $\Phi^*(x_2)\,\Phi(x_2)$ in H_1 can make up the closed loop).

By making the insertion we have distinguished between the two particle lines, and therefore gain a factor of 2.

Consider a more complicated graph such as that shown below. After the first interaction on the bottom,

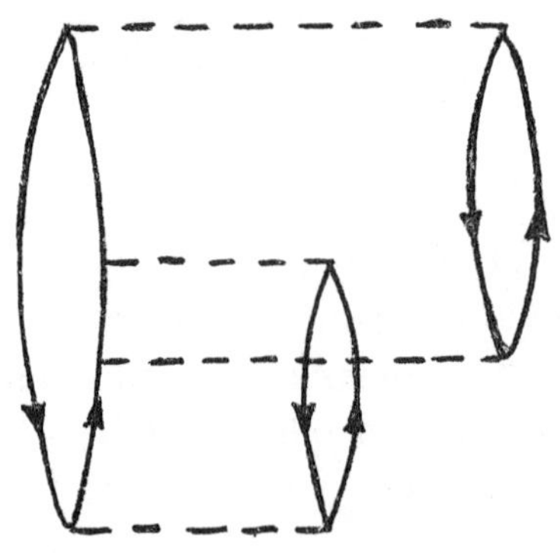

the second one distinguishes, on the left, between two particle lines, since it could have begun from either of the two left-hand particle lines; the next interaction distinguishes between two hole lines, etc. The net numerical factor is 1 since there are no equivalent lines. This rule greatly simplifies the deduction of the numerical factor.

We shall now show, by a simple physical argument, why the coefficient of -i t is the correction to the energy, before going on to more rigorous arguments in the next section.

Let us consider the perturbing Hamiltonian to be time-dependent, so that it is zero at $t = 0$, and the system is in state Φ_0. We then turn the perturbation on slowly and smoothly, so that the system is carried up to the ground state Ψ_0 of the perturbed system, which has energy E_0. After a long time T, the perturbation is turned off slowly and smoothly. Let us now look at the matrix element

$$(\Phi_0, \hat{U}(t)\, \Phi_0) = e^{iW_0 t}\, (\Phi_0, U(t)\, \Phi_0) \qquad \text{I(23)}$$

for $t = T + 2\Delta$, where Δ is the time taken to turn the perturbation on and off. According to our argument,

$$U(\Delta)\, \Phi_0(0) = \Psi_0(\Delta), \qquad \text{I(23.1)}$$

that is, the development operator U leads the ground state of the unperturbed system into that of the perturbed one in time Δ. Then,

$$U(T + \Delta)\, \Psi_0(\Delta) = e^{-iE_0 T}\, \Psi_0(\Delta) \qquad \text{I(23.2)}$$

since $U(t) = \exp(-iHt)$. Between the time $T + \Delta$ and $T + 2\Delta$, the system is then led back down to the unperturbed ground state by turning the perturbation off.

As $T \to \infty$, we can neglect the contributions to the phase of (23) from the two times Δ, so that the matrix element (23) goes as

$$e^{iW_0 t}(\Phi_0, U(t)\Phi_0) \underset{t \to \infty}{\to} e^{i(W_0 - E_0)t} \qquad \text{I}(24)$$

Our work above has been to expand (24), which is the same as (12), in a power series. If we let

$$E_0 = W_0 + W^{(1)} + W^{(2)} + \cdots$$

then (24) is just

$$e^{-i(W^{(1)} + W^{(2)} + \cdots)t}, \quad t \to \infty \qquad \text{I}(25)$$

and what we have done in this section is to expand the exponent out. Thus, the result of the second-order pairing illustrated in Fig. 3 was just to reproduce $-iW^{(2)}t$, as noted after (21). We see, therefore, why the term depends linearly on t. This argument is adequate for obtaining the correction to the energy which we get from the phase of (23). Since the phase goes to ∞ linearly with t, the finite changes we make in the problem by turning on and off the interaction do not matter. However, if one wishes to look at parts of (23) which remain finite, one should be more careful. We shall discuss this in connection with the renormalization of the wave function in the next section.

We see, further, that if we continued to expand

(25) out, we would have a term $(-iW^{(2)}t)^2$, which must look graphically like

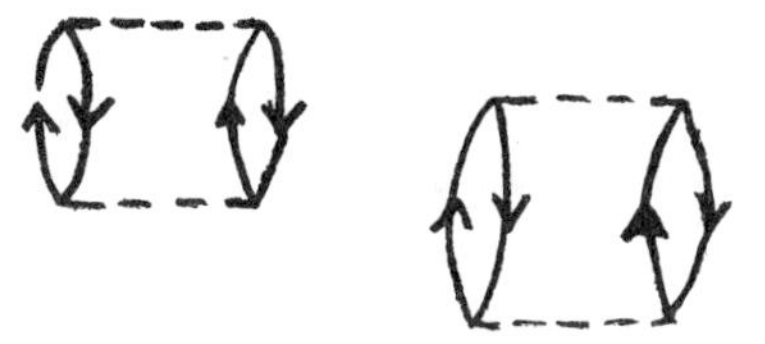

It is clear, further, that this term depends on the total number of particles N as N^2, since $W^{(2)} \propto N$. Hence our expansion of (12), which is equivalent to that of (25), will seem highly divergent. It is clear from the above that we must simply collect terms in the expansion of (12) so as to regain the exponential form (25), and this is what we shall proceed to do in the next section. It is already clear from our simple arguments, however, that we must just add up the linked graphs, i.e., catch each $W^{(n)}$ the first time it appears in the expansion of (25), in order to obtain the correction to the energy.

II. EXCITATIONS AND GREEN'S FUNCTIONS

A. Single-Particle Excitations

In this section we shall introduce Green's Functions for systems at zero temperature, that is, systems in which the N particles fill the first N levels in the ground state. At first sight, the study of excited states of such a system seems to present unsurmountable difficulties. Already in a medium-weight nucleus, there are 10^6 levels per MeV at excitation energies of 8-10 MeV. Extended systems, such as solids and liquids, usually have excited states going continuously in energy either down to the ground state, or to the "energy gap".

Fortunately, a complete knowledge of the excited states of a complicated system is not required. Many of the properties of interest can be calculated from knowledge of the single-particle excitations, i.e., from those states reached by transferring one particle from below to above the Fermi energy. For example, the electric polarizability can be obtained from the quantity

$$\sum_n \frac{|V_{on}|^2}{E_o - E_n}$$

where V_{on} is the matrix element of the applied electrostatic potential between the ground state of the system and the excited state of energy E_n. The electrostatic potential is a single-particle operator, and has matrix elements only between configurations which differ in the level of a single particle.

There is one deep point, which one should understand, but which does not turn out to be important in practical calculations. This is that the states reached by transferring one particle from below to above the Fermi level, are not eigenstates of the system and, indeed, have, in general, only a very small overlap with any given eigenstate. The single-particle excitation will be mixed into many neighbouring eigenstates, each of which will have more complicated components consisting of two-, three-, four- etc.

particle excitation. However, in calculating quantities such as the polarizability, which depend only on the single-particle component, we can use the single-particle excitation spectrum. Strictly speaking we should give the energy E_n a "half-width" $-i\Gamma/2$, where Γ is the distance over which the single-particle excitation is mixed into eigenstates of the system.

We shall call those states reached in transferring one or two particles "excitations" to distinguish them from the eigenstates.

A convenient description of the excitations of a system can be given in terms of Green's Functions which we shall now introduce, following especially the treatment of Galitskii and Migdal, Zh. eksper, teor. Fiz. **34**, 139 (1959).

B. The One-Particle Green's Function

The single-particle Green's Function is defined in momentum representation as

$$G(k_1 t_2 - t_1) \equiv i\langle 0 | T[a_k(t_2) a_k^+(t_1)] | 0\rangle \qquad \text{II}(1)$$

where $|0\rangle$ is the true ground state of the N-particle system (which we called Ψ_0 before), and where $a_k(t)$ is now the operator a_k in the Heisenberg representation,

$$\text{and } a_k(t) = e^{iHt} a_k e^{-iHt}. \qquad \text{II}(1.1)$$

Now $H|0\rangle = E_o|0\rangle$ so that*)

$$G(k,t_2 - t_1) = i\langle 0 | a_k e^{-i(H-E_0)(t_2 - t_1)} a_k^+ | 0\rangle, t_2 > t_1$$

$$= -i\langle 0|a_k^+ e^{i(H-E_0)(t_2-t_1)} a_k|0\rangle, \; t_2 < t_1$$

II(2)

In the case of free, non-interacting particles, we see that

$$G^{(0)}(k, t_2 - t_1) = ie^{-iT_m(t_2-t_1)}, \; t_2 > t_1$$

$$= -ie^{-iT_i(t_2-t_1)}, t_2 < t_1$$

II(2.1)

where T_m and T_i are the kinetic energies. These Green's Functions were just the insertions that we were required to make following Wick's theorem on pairing. One can easily include an external potential in (2.1) by letting $T_m \rightarrow T_m + V_m$.

Let $t_2 - t_1 = \tau$. Then,

$$\lim_{\tau \rightarrow -0} G(k, \tau) = -i\langle 0|a_k^+ a_k|0\rangle = -in_k. \qquad \text{II(3)}$$

and

$$\langle o|T|o\rangle = i\,\Sigma \frac{k^2}{2M} \lim_{\tau \rightarrow -0} G(k, \tau). \qquad \text{II(3.1)}$$

To calculate the potential energy, take

$$\lim_{\tau \rightarrow -0} \frac{\partial G(k, \tau)}{\partial \tau} = \langle o|a_k^+ (H-E_o) a_k|o\rangle. \qquad \text{II(4)}$$

*) These two parts of the Green's Function are often called the retarded and advanced parts, respectively.

Using, as before,

$$H = \sum_k (k|T|k)\, a_k^+ a_k + \frac{1}{2} \sum_{k_1 k_2 k_3 k_4} (k_1 k_2|V|k_3 k_4)\, a_{k_2}^+ a_{k_1}^+ a_{k_3} a_{k_4}$$

and

$$(H-E_o)\, a_k = a_k\, (H-E_o) + [H, a_k]$$

and lettii.ꞈ H operate on the ground state, we see that

$$\lim_{\tau \to -0} \frac{\partial G(k,\tau)}{\partial \tau} = \langle o|a_k^+ [H, a_k]|o\rangle$$

$$= - \langle o|(k|T|k) a_k^+ a_k + \sum_{k_1 k_3 k_4} (k_1 k|V|k_3 k_4)\, a_k^+ a_{k_1}^+ a_{k_3} a_{k_4} |o\rangle \qquad \text{II(5)}$$

so that $\sum_k \lim_{\tau \to -0} \frac{\partial G(k,\tau)}{\partial \tau}$

$$= - \langle o|T|o\rangle - 2\langle o|V|o\rangle \qquad \text{II(5.1)}$$

Consequently, knowing $G(k, \tau)$ and $\partial G/\partial \tau$, we can find the expectation values of kinetic and potential energies in the ground state.

We shall now obtain a spectral representation of G. Define

$$G(k, E) = \int_{-\infty}^{\infty} e^{iEt}\, G(k, t)\, dt$$

$$= \langle o| a_k \frac{1}{H - E_o - E - i\delta} a_k^+ |o\rangle \qquad \text{II}(6)$$

$$- \langle o| a_k^+ \frac{1}{H - E_o + E - i\delta} a_k |o\rangle$$

$$= G_1(k, E) + G_2(k, E)$$

We have considered E to have a small positive imaginary part for $t > 0$ and a small negative imaginary part for $t < 0$, making the oscillatory parts in the above integral cut down at $+\infty$ and $-\infty$. Otherwise, the integral would be undefined. We now write the intermediate states in the energy representation, assuming that we have a large, but finite system,

$$G_1 = \sum_p |\langle \Psi_p^{N+1} | a_k^+ | \Psi_o^N \rangle|^2 \frac{1}{W_p - E_o - E - i\delta} \qquad \text{II}(7)$$

$$G_2 = -\sum_p |\langle \Psi_p^{N-1} | a_k | \Psi_o^N \rangle|^2 \frac{1}{W_p' - E_o + E - i\delta}$$

where we have specifically written out the fact that G_1 involves the matrix elements of a_k^+ between the ground state for N particles and the states of N + 1 particles. The matrix elements in (7) do not depend upon E, so that they will not affect the analytic properties of G as a function of E.

We see immediately that G_1 has poles slightly below the real axis at $E = W_p - (E_o + i\delta)$, so that knowledge of the poles of G_1 defines the states of the $(N + 1)$-particle system. The poles of G_2 lie above the real axis at $E = -(W_p^1 - E_o - i\delta)$. Introducing the chemical potential μ by

$$\mu = W_o - E_o$$

$$\mu' = E_o - W'_o$$

Then the poles in G_1 lie near the real axis above μ, and those in G_2 lie below μ'. For a large system, and no pairing forces, $\mu \cong \mu'$.

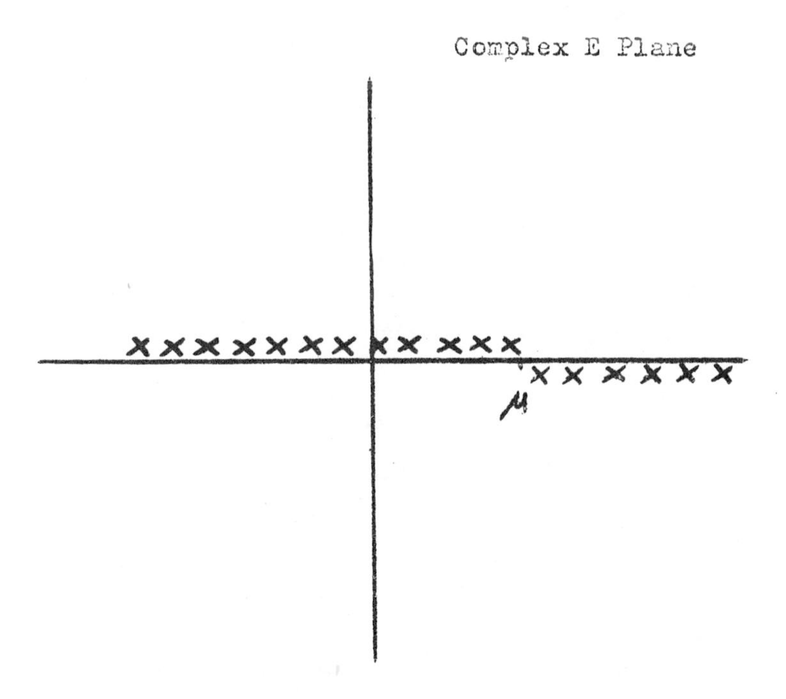

Fig. 9. Poles of G(k, E) for a Finite System

If we now wish to consider the properties of G(k, E) for complex values of E, we note that the retarded part of G(k, E) can be defined from (6) only for values of E in the upper half plane, and the advanced part, only for those in the lower half plane. However, the second equation in (6), or eq. (7), allows the functions to be analytically continued into the other half plane. In particular, we have assumed that the system is finite, so that the poles are discrete, and we can "sneak" between them. It is clear from the definition of G that

$$G(k, E^*) = G^*(k, E) \qquad \text{II(8)}$$

As the number of particles increases while the density remains constant, the eigenvalues of the Hamiltonian get closer together. Although the singularities of G(k, E) become indefinitely close together, the function remains analytic off the real axis even for an infinite system. One can then define its values on the upper and lower sides of the real axis as limits of its values in the upper and lower half-planes.

When the energy levels of the system are very close together, it is convenient to define

$$g_I(W)\,\delta W = \sum_{W < W_p < W + \delta W} |\langle \Psi_p^{N+1} | a_k^+ | \Psi_o^N \rangle|^2 = |\langle \Psi_p^{N+1} | a_k^+ | \Psi_o^N \rangle|^2 \frac{\delta W}{D} \qquad \text{II(9)}$$

where D is the average spacing, so that $\delta W/D$ represents the number of levels in the energy interval δW.

Then, off the real axis, we can define the Green's function as

$$G(k, E) = \int_0^\infty \frac{dW\, g_I(W + W_o)}{W + \mu - E - i\delta} - \int_0^\infty \frac{dW\, g_{II}(W + W'_o)}{W - \mu + E - i\delta}$$

$$\omega = W - W_o \qquad\qquad \omega = W' - W''_o$$

$$= \int_0^\infty \frac{d\omega A_+(k, \omega)}{\omega + \mu - E - i\delta} - \int_0^\infty \frac{dw A_-(k, \omega)}{\omega - \mu + E - i\delta} \,,$$

$$\begin{aligned} A_+(k, \omega) &= g_I(\omega + W_o) \\ A_-(k, \omega) &= g_{II}(\omega + W'_o) \end{aligned} \qquad \text{II(10)}$$

where g_{II} is defined analogously.

The matrix element $<\Psi_p^{N+1}|a_k^+|\Psi_o^N>$ has a simple physical meaning. The Ψ_o^N is the physical interacting ground state; the a_k^+ then creates a particle in the single-particle state k. This matrix element is, therefore, the amplitude with which the N particles in their ground state and the single particle in state k is to be found in the (N + 1)-particle eigenstate p. The quantity $g_1(W)$ is called the "strength function." In fact, the quantity $|<\Psi_p^{N+1}|a_k^+|\Psi_o^N>|^2$ is essentially what is called the "reduced width" in nuclear physics; it is the probability of finding the single particle and unexcited core in compound state p. By working with an infinite system, we have avoided having to bring the particle k in through a centrifugal or Coulomb barrier, and have, consequently, sidestepped the complications of introducing penetrabilities which enter into problems of finite nuclei.

These matrix elements clearly obey the sum rule

$$\sum_p |\langle\Psi_p^{N+1}|a_k^+|\Psi_o^N\rangle|^2 + \sum_p |\langle\Psi_p^{N-1}|a_k|\Psi_o\rangle|^2 =$$

$$\langle\Psi_o|a_k a_k^+ + a_k^+ a_k|\Psi_o\rangle = 1 \qquad \text{II}(11)$$

This sum rule is similar to the one used by Lane, Thomas and Wigner (Phys. Rev. **98**, 693 (1955)) in nuclear studies. Namely, they show that if one adds a particle in a given single-particle state to the nucleus, then all of this "excitation" is to be found somewhere in compound states, e.g.

$$\sum_p |\langle\Psi_p^{N+1}|a_m^+|\Psi_o^N\rangle|^2 = 1 \qquad \text{II}(11.1)$$

They neglect the antisymmetrization of the particle in state m with those in Ψ_o^N, which amounts to saying that one has A particles of one kind and then adds a particle of another kind. In this case, $a_m|\Psi_o^N\rangle = 0$, so that (11.1) is identical with (11) in the case where m is a distinguishable particle; e.g., in the case where one adds a meson to the nucleus, and looks for the compound states of the A nucleons and one meson.

In the actual case, one should antisymmetrize of course. This can lead to important corrections, since in the interacting ground state $|\Psi_o^N\rangle$ the state k in (11) may be occupied by one or another of the particles with appreciable probability, so that one needs the second term on the left-hand-side of (11) to complete the sum rule.

From eq. (10) we note that, since g_I and g_{II} are real, contributions to the imaginary part of G(k, E) come only from the regions of the poles in the denominators. Presence of the δ's in the denominators tells us how to go around these poles. Calculating this imaginary part, we find

$$\begin{aligned} \text{Im } G(k, E) &= i\pi\, g_{I}(E),\ E > \mu \\ &= -\, i\pi\, g_{II}(E),\ E < \mu' \end{aligned} \qquad \text{II}(11.2)$$

where we have taken the ground-state energy E_0 to be zero for convenience. The sum rule eq. (11) can then be reexpressed as

$$\int_{\mu}^{\infty} \text{Im } G(k, E)\, dE - \int_{-\infty}^{\mu} \text{Im } G(k, E)\, dE = i\pi \qquad \text{II}(11.3)$$

where we have set $\mu = \mu'$.

C. Perturbation Calculation of Green's Functions

We can calculate the Green's Functions by the graphical methods introduced earlier. Let us calculate the quantity

$$\langle \Phi_0 | U(T_2 - t_2) a_k U(t_2 - t_1) a_k^+ U(t_1 - T_1) | \Phi_0 \rangle$$

where Φ_0 is the unperturbed ground state and where $T_2 > t_2 > t_1 > T_1$; for this we can use Wick's theorem. Firstly, however, we must convert the U's into $\hat{U}$'s, which have the straightforward expansion (17.1). We first insert factors of $\exp(\pm iH_0 t)$ so as to make a_k and a_k^+ into $a_k(t)$ and $a_k^+(t)$, the time-dependent operators in the interaction representation. Then the above quantity becomes

$$<\Phi_o|U(T_2 - t_2)e^{-iH_ot_2}a_k(t_2)e^{iH_ot_2}U(t_2 - t_1)e^{-iH_ot_1}$$

$$\times \; a_k^+(t_1)e^{iH_ot_1}U(t_1 - T_1)|\Phi_o>.$$

Now, from eq. I(3.4),

$$e^{iH_ot_2}U(t_2 - t_1)e^{-iH_ot_1} = \hat{U}(t_2, t_1)$$

so that

$$<\Phi_o|U(T_2 - t_2)a_kU(t_2 - t_1)a_k^+U(t_1 - T_1)|\Phi_o> =$$

$$<\Phi_o|\{\hat{U}(T_2, t_2)a_k(t_2)\hat{U}(t_2, t_1)a_k^+(t_1)\hat{U}(t_1, T_1)\}|\Phi_o> e^{-iW_o(T_2-T_1)}.$$

Noting that the operators occur in a time-ordered sequence, we see, using I(7.1) that the quantity in brackets is

$$\{\,\} = \sum_{n=o}^{\infty}\frac{(-i)^n}{n!}\int_{T_1}^{T_2} - - - \int_{T_1}^{T_2} T[H_1(t'_n) - - - H_1(t'_2)$$

$$H_1(t'_1) \quad \times a_k(t_2)a_k^+(t_1)]dt'_n - - - dt'_1$$

which is now in the form suitable for drawing graphs. We have to draw a line k entering the graph at time t_1, and a line k leaving the graph at t_2, to represent a_k^+ and a_k, respectively. The perturbation expansion of the development operators is represented by any number of vertices between T_1 and T_2 so that the process of contraction gives graphs like the groundstate ones, except that there is now an external line. Some of the graphs are shown below in fig. 10. The graphs can be either

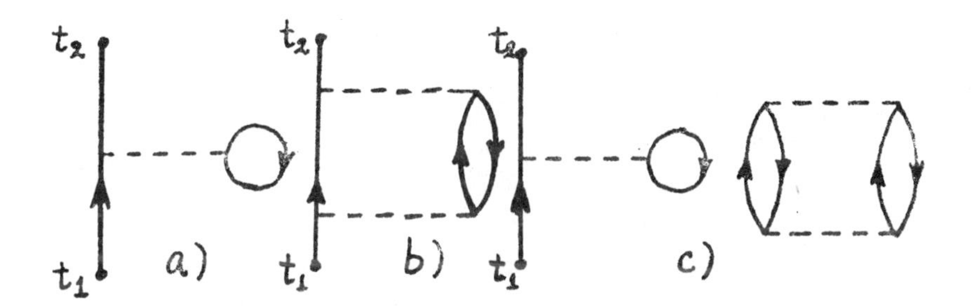

Fig. 10a) and b) are examples of linked graphs, whereas c) is an unlinked one. The dots indicate the points at which the operators a_k^+ and a_k operate.

linked or unlinked, the unlinked ones giving just

$$<\Phi_o|\hat{u}(T_2 - T_1)|\Phi_o>.$$

Consequently, the sum of linked graphs with two external lines gives

$$\frac{<\Phi_o|U(T_2 - t_2)a_k U(t_2 - t_1)a_k^+ U(t_1 - T_1)|\Phi_o>}{<\Phi_o|U(T_2 - T_1)|\Phi_o>}$$

We set

$$\langle\Phi_o| = \sum_n \langle\Phi_o|\Psi_n\rangle\langle\Psi_n|$$

and let the imaginary parts of $-T_1$ and T_2 tend to $-\infty$. This is just the prescription needed to get rid of oscillatory terms such as in eq. I(21). The above fraction then tends to

$$\frac{1}{i} G(k, t_2 - t_1) = \langle\Psi_o|e^{iHt_2} a_k e^{-iH(t_2-t_1)} a_k^+ e^{-iHt_1}|\Psi_o\rangle$$

which is just the retarded part of the Green's Function. The advanced part can be calculated in the same way, except that the external line must enter the graph above the point at which it leaves.

We can now use our usual graphical techniques to calculate the Green's Function. Let us consider, by way of example, the Green's Function describing propagation with any number of lowest-order "self-energy" insertions, that is, the Green's Function describing the processes shown in fig. 11.

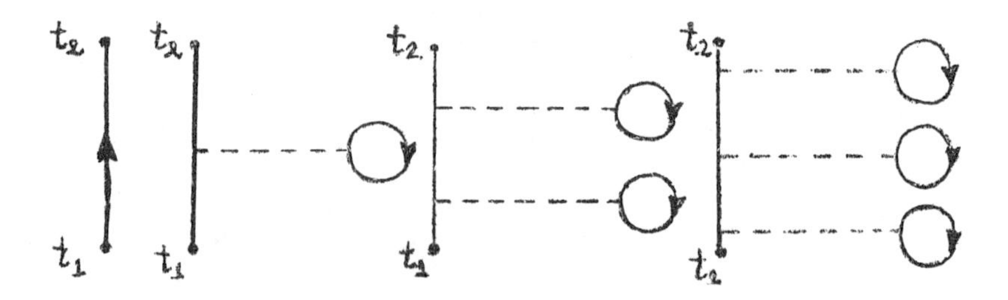

Fig. 11. Green's Function with any number of lowest-order self-energy insertions.

Here the free propagator, $(T_k - E)^{-1}$, which we represent graphically by ↑

is modified by the factor $(1 + x + x^2 + - - -)$ where x is the quantity defined in section II. Changing notation, we now call

$$x = \frac{\Sigma_1(k)}{E - T_k}, \quad \Sigma_1(k) \equiv \sum_l \{(kl|V|kl) - (kl|V|lk)\} . \qquad \text{II}(12)$$

The Green's function is then

$$G(k, E) = \frac{1}{T_k - E} - \frac{1}{T_k - E}\Sigma_1(k)\frac{1}{T_k - E}$$

$$+ \frac{1}{T_k - E}\Sigma_1(k)\frac{1}{T_k - E}\Sigma_1(k)\frac{1}{T_k - E} - - \qquad \text{II}(13)$$

$$= \frac{1}{T_k + \Sigma_1(k) - E},$$

One can calculate Σ to higher order; for example, we can include second-order terms such as shown in fig. 12, which now depend on E and still sum them, to

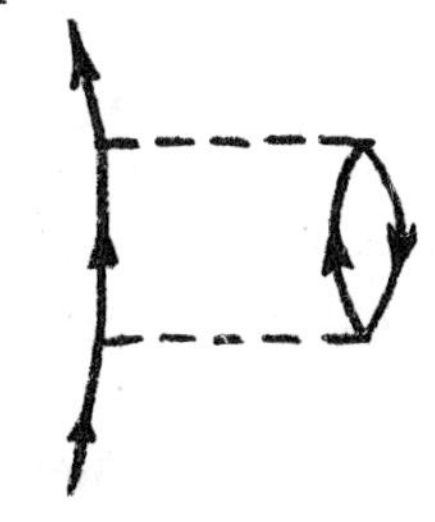

Fig. 12. A second-order self-energy insertion.

give (12) with $\Sigma_1(k)$ replaced by $\Sigma(k, E)$, where Σ now includes the other insertion.

Eq. (13) is easily seen to be equivalent to the integral equation

$$G(k, E) = \frac{1}{T_k - E} - \frac{1}{T_k - E} \Sigma(k, E) G(k, E) . \qquad \text{II}(14)$$

Going back to the time-representation, using

$$\Sigma(k, t) = \frac{1}{2\pi} \int_{-\infty}^{\infty} e^{-iEt} \Sigma(k, E)\, dE \qquad \text{II}(15)$$

we find that (14) becomes an integral equation

$$G(k, t) = G^{(0)}(k, \tau)$$

$$- \int_{-\infty}^{\infty} d\tau' \int_{-\infty}^{\infty} d\tau'' G^{(0)}(k, \tau - \tau'') \Sigma(k, \tau' - \tau'') G(k, \tau'') \qquad \text{II}(16)$$

where

$$G^{(0)}(k, \tau) = \frac{1}{2\pi} \int_{-\infty}^{\infty} \frac{e^{-iE\tau}}{T_k - E} dE$$

is given in eq. (2.1).

Eq. (16) has a simple interpretation. Eq. (15) indicates that the particle can either propagate (through $G^{(0)}$) without interaction, or it can have any number of interactions (through G) up to the time τ'' at which time it has one final interaction (through Σ) and propagates without interaction to time τ. Such integral equations were first written down by Dyson in connection with the self-energy problem in field theory.

In the Hartree-Fock approximation given by (13)

it is seen from (12) that, since Σ is real, the poles in G lie on the real axis.

In the next approximation, we include processes like those in fig. 13.

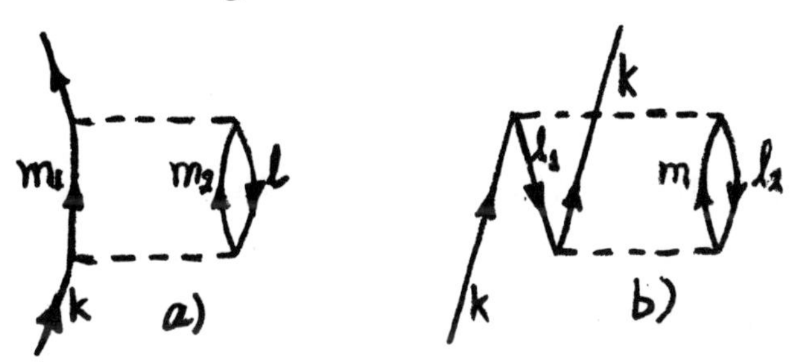

Fig. 13. Processes which enter into the renormalization of the pole.

and their exchange terms in Σ, i.e.

$$\Sigma_2 = \frac{1}{2} \sum_{l, m_1, m_2} \frac{|(kl|V|m_1m_2) - (kl|V|m_2m_1)|^2}{E + T_l - T_{m_1} - T_{m_2}}$$

$$+ \frac{1}{2} \sum_{l_1, l_2, m} \frac{|(l_1l_2|V|km) - (l_1l_2|V|mk)|^2}{E + T_m - T_{l_1} - T_{l_2}} \qquad \text{II}(17)$$

The inclusion of this term now changes the situation essentially, since E appears in Σ_2

Let us now investigate the poles in

$$G_2(k, E) = \frac{1}{T_k + \Sigma_1(k) + \Sigma_2(k, E) - E} \, . \qquad \text{II}(18)$$

One of the poles, and the one we shall be mainly interested in, is at energy ϵ, given by

$$T_k + \Sigma_1(k) + \Sigma_2(k, \epsilon) = \epsilon. \qquad \text{II}(18.1)$$

Expand

$$\Sigma_2(k, E) = \Sigma_2(k, \epsilon) + \Sigma_2'(k, \epsilon)(E - \epsilon) + \cdots \qquad \text{II}(18.2)$$

For E in the neighbourhood of ϵ, then

$$G(k, E) \cong \frac{1}{(\epsilon - E) + (E - \epsilon)\Sigma_2'} \qquad \text{II}(18.3)$$

so that the residue at the pole is

$$Z_k = \frac{1}{1 - \Sigma_2'(k, \epsilon)}. \qquad \text{II}(18.4)$$

There is a simple physical interpretation of the renormalization through $d\Sigma/dE$. Carrying out the differentiation, we find

$$\frac{d\Sigma_2}{dE} = -\frac{1}{2}\sum_{l, m_1, m_2}\left[\frac{|(kl|V|m_1m_2) - (kl|V|m_2m_1)|}{E + T_l - T_{m_1} - T_{m_2}}\right]^2 - \frac{1}{2}\sum_{l_1, l_2, m}\left[\frac{|(l_1l_2|V|km) - (l_1l_2|V|mk)|}{T_{l_1} + T_{l_2} - E - T_m}\right]^2. \qquad \text{II}(18.5)$$

The first term in (18.5) shown in Fig. 13a, gives the probability that the single-particle excitation k will be mixed into the two-particle, one-hole excitation m_1, m_2, l. If the potential V contains a strong short-range interaction, then the matrix element $|(kl|V|m_1m_2) - (kl|V|m_2m_1)|$ will decrease only slowly as the energy of the two-particle, one hole excitation, equal to $T_{m1} + T_{m2} - T_l$, moves away from the pole. In this case, bits and pieces of the single-particle excitation k will be mixed far and wide into two-particle, one hole excitations. The second term in (18.5) describes the fact that two particles of the N-particle system, originally in states l_1 and l_2, can interact, going to states k and m. This makes it then impossible to put the $(N + 1)^{th}$ particle into state k. It can be seen from (18.4) and (18.5) that both of these processes reduce the residue at the pole, as one might think from the above physical interpretation of the terms. The sum rule, eq. (11.3), indicates that the bits and pieces of strength not in the above Z_k must be found elsewhere in the spectrum; thus, there are other poles in G corresponding to more complicated states (e.g., two-particle, one-hole as in fig. 13a) where these bits occur.

We next discuss the position of the pole in G_2 of eq. (18). The real part of the denominator of the first term on the right-hand-side of (17) can go to zero, so that Σ_2 can have an imaginary part. (Definition of the integral (13) requires giving E a small positive imaginary part, so that how to go around the pole is well defined.) Considering the sum over m_1 and m_2 for fixed l, we see that we need sum over only m_1, since from momentum conservation $m_1 + m_2 = k + l$.

One sees that the residue is negative, and proportional to the value of $|(kl|V|m_1m_2) - (kl|V|m_2m_1)|^2$ at the point of energy conservation $E + T_l = T_{m1} + T_{m2}$. This describes the decay of the single-particle excitation into two-particle, one-hole excitations at the same energy. That this describes a decay can be seen from (18), since the negative imaginary part of Σ_2 gives E a negative imaginary part at the pole.

D. Quasiparticles

In the last section, we found that if we calculate the Green's Function by perturbation theory, we obtain an expression

$$G_I(k, E) = \frac{1}{T_k + \Sigma(k, E) - E - i\delta} \qquad \text{II}(19)$$

where G_I is the part of the Green's Function describing particle (in distinction to hole) excitations.

From eq. (11.2), we see that the strength function $\pi g_I(E)$ can be obtained from the imaginary part of G_I, giving us the following expression

$$\pi g_I(E) = \frac{-\mathrm{Im}\Sigma(k, E)}{(T_k + \mathrm{Re}\Sigma(k, E) - E)^2 + (\mathrm{Im}\Sigma(k, E))^2}. \qquad \text{II}(19.2)$$

Comparing this with the expression for noninteracting particles where

$$G^{(0)}(k, E) = \frac{1}{T_k - E - i\delta} \qquad \text{II}(19.3)$$

$$\pi g_I^{(0)}(k,E) = \pi\,\delta(T_k - E) \qquad \text{II}(19.4)$$

we see that the interactions have spread the strength in (19.2) over a region of width $|\text{Im}\,\Sigma(k,E)|$ in energy.

In the neighbourhood of the pole discussed in the last section, G has the behaviour

$$G_I(k,E) \cong \frac{Z_k}{T_k + \Sigma(k,\epsilon) - E'} \qquad \text{II}(19.5)$$

which, aside from the factor Z_k, will give a strength function similar to that in eq. (19.2) in the region where g_I is large in cases of interest where $\Sigma(k,\epsilon)$ does not differ much from $\Sigma(k,E)$.

The Fourier transform of (19.5) is

$$G_I(k,\tau) = Z_k\, e^{i[\epsilon_k - i\Gamma(k)/2]\tau} \qquad \text{II}(19.6)$$

where $\epsilon_k = T_k + \text{Re}\Sigma(k,\epsilon)$, $\Gamma(k)/2 = -\text{Im}\Sigma(k,\epsilon)$.

This Green's Function has the same form as for an independent particle, all many-body features of the problem having gone into the ϵ_k and $\Gamma(k)$, the latter giving this excitation a half life $\Gamma(k)^{-1}$.

Clearly, such a description will be useful only if Z_k is not small compared with unity, so that an appreciable fraction of the single-particle strength remains in this excitation, and if $\Gamma(k)$ is not too large, so that the excitation has a reasonably long lifetime. The size of Z_k depends, of course, on the strength of interaction; strong interactions will mix the single-particle strength widely into more complicated states. Even for fairly strong interaction, however, $\Gamma(k)$ can be made small by letting k approach k_F. For k near k_F, one finds* from evaluating the imaginary part of Σ_2, eq. (17), for $E = \epsilon_k$ that $\Gamma(k)$ goes as

$$\Gamma(k) \cong C(k - k_F)^2 .$$

This behaviour follows directly from the Pauli principle and phase-space arguments. In eq. (17), one can sum freely over l and m_1, m_2 then being fixed by momentum conservation. The imaginary part comes from the pole at

$$\epsilon_k = T_{m_1} + T_{m_2} - T_l .$$

From this, one sees that none of the momenta can differ in magnitude more than $\sim |k - k_F|$ from k_F, although there is no restriction to particular angular regions. Thus, the phase-space for each of the two free sums is proportional to $|k - k_F|$.

In the region of the Fermi momentum k_F, then, the excitations are long-lived and their lifetime becomes infinite as $k \to k_F$.

*As noted earlier, evaluation of the sum in eq. (17) can be carried out by converting it to an integral, in which case the poles can be handled by giving E in the denominator a small positive imaginary part.

These considerations apply to systems that can be described, as above, either in perturbation theory or in theories obtained by making partial summations in perturbation theory. Such systems are called normal systems. It is not clear how to decide whether a system is normal or not, a priori; one just has to see how to make a description of each system in turn.

In normal systems there is an excitation of the above type corresponding to each momentum k; that is, there is a one-to-one correspondence between the excitations in the perturbed and the unperturbed system. We call the excitations in the perturbed system "quasiparticles." The quasiparticles behave very much like ordinary particles, except that the dependence of their energy ϵ_k on momentum k is more complicated and they have a lifetime and a reduced amplitude Z_k.

E. Time-Dependent Methods

We have calculated the expression for G_I, the part of G describing particle propagation. Were we to calculate G_{II}, the part describing hole propagation, we would obtain the same expression, but with changed sign of δ. We can thus say

$$G(k,E) = \frac{1}{T_k + \Sigma(k,E) - E - i\delta \operatorname{sgn}(E - E_F)}$$

We now discuss the calculation of $\Sigma(k,E)$. From the graph, Fig. (14) one finds

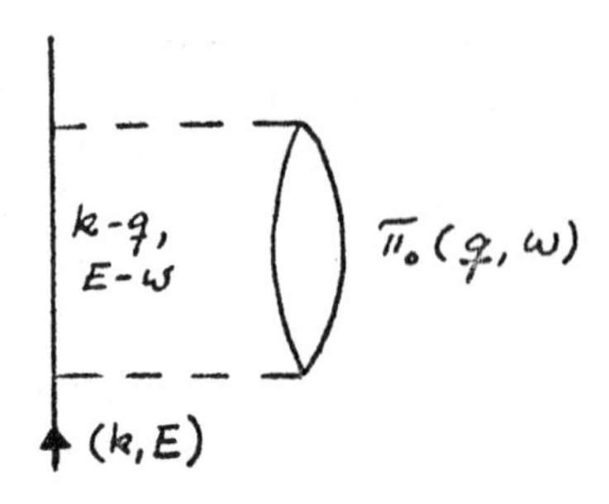

Fig. 14. Second-order self-energy insertion.

$$\Sigma(k,E) = \frac{1}{(2\pi)^4} \int d^3q \int d\omega \, [-iG(k-q, E-\omega)] \qquad \text{II(20)}$$

$$[-iV_q]^2 \, [-i\pi_0(q,\omega)]$$

where V_q is the Fourier transform of the potential. Here $\pi_0(q,\omega)$ is obviously made up out of two Green's functions

$$-i\,\pi_0(q,\omega) = \frac{-2}{(2\pi)^4} \int d^3p \int dE \, (-i)^2 G(p,E)$$

$$G(p-q, E-\omega) \qquad \text{II(20.1)}$$

Let us now evaluate $\pi_0(q,\omega)$ assuming that the quasi-particle approximation is good; i.e., replacing $T_k + \Sigma(k,E)$ by ϵ_k and forgetting about the renormalization factors R_k. If both G's refer to particles

$$(T_k > k_F^2/2m)$$

or if both refer to holes, then we can complete the contour in the ω integration by a semicircle in the upper or lower half plane, respectively, such that the contour encloses no poles, and therefore these contributions vanish. Assume now that $p > k_F$, $p - q < k_F$. If we close the contour in the lower half plane, we pick up the residue at $E = \epsilon_p - i\delta$ giving

$$[\pi_0(q,\omega)]_1 = \frac{4\pi}{(2\pi)^4} \int_{\substack{p < k_F \\ |p - q| < k_F}} d^3p \; \frac{1}{\epsilon_p - \epsilon_{p-q} - \omega - i\delta} \qquad \text{II(20.2)}$$

For $p < k_F$, $p - q > k_F$ we get the contribution

$$[\pi_0(q,\omega)]_2 = \frac{-4\pi}{(2\pi)^4} \int_{\substack{p > k_F \\ |p - q| > k_F}} d^3p \; \frac{1}{\epsilon_p - \epsilon_{p-q} - \omega + i\delta} \qquad \text{II(20.3)}$$

In terms of the time-ordered graphs, $[\pi_0]_1$ corresponds to the bubble in Fig. 15 a; $[\pi_0]_2$ to the bubble in Fig. 15 b.

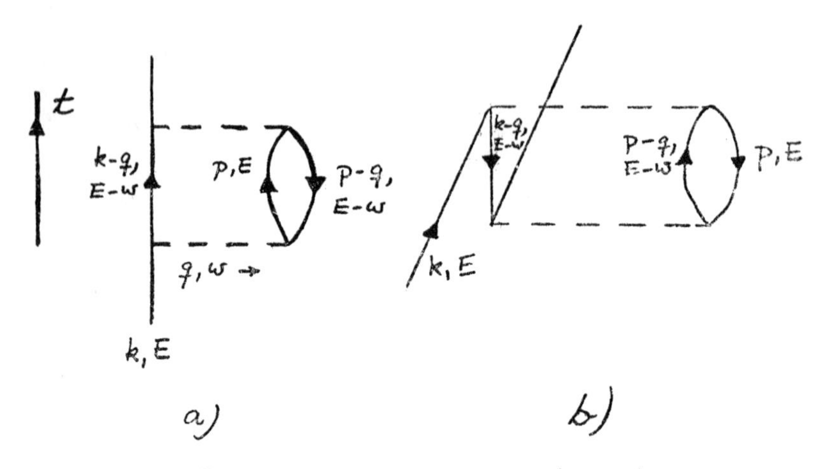

Fig. 15. Graphs showing how $\pi_0(q - \omega)$ enters into the second-order self-energy.

III. GROUND-STATE ENERGY THEOREMS AND FUNCTIONAL DIFFERENTIATION OF THE ENERGY

From equations II(4) and following, we know that

$$\frac{dG(k,t)}{dt}\bigg|_{t=-0} = \sum_k (0 \mid a_k^+ [H,a_k] \mid 0)$$

$$= - (0 \mid T + 2V \mid 0) \qquad \text{III(1)}$$

$$= \sum_n (W'_n - E_0) \sum_k \langle 0 | a_k^+ | n \rangle \langle n | a_k | 0)$$

Now

$$(W'_n - E_0) = (W'_n - W'_0) + (W'_0 - E_0) = \omega_{n0} - \mu \qquad \text{III(1.1)}$$

so that

$$(0 | a_k^+ \, [H, a_k] \, | 0) = \int_0^\infty A_-(k,\omega)(\omega - \mu) \, d\omega \qquad \text{III(1.2)}$$

Hence,

$$\left.\frac{dG(k,t)}{dt}\right|_{t=-0} = \int_0^\infty A_-(k,\omega)(\omega - \mu) \, d\omega \qquad \text{III(1.3)}$$

Now

$$G(k,\omega) = \int_0^\infty d\omega' \left\{ \frac{A_+(k,\omega')}{\omega' - \omega + \mu - i\delta} - \frac{A_-(k,\omega')}{\omega' + \omega - \mu - i\delta} \right\} \qquad \text{III(1.4)}$$

and

$$\frac{1}{2\pi i} \int_C G(k,\omega) \, \omega d\omega = - \frac{1}{2\pi i} \int_0^\infty d\omega' \int_C d\omega \, \frac{A_-(k,\omega')\,\omega}{\omega' + \omega - \mu - i\delta}$$

$$\int_0^\infty d\omega' \, A_-(k,\omega')(\omega' - \mu) \qquad \text{III(2)}$$

where C is a contour consisting of the real axis and a semicircle enclosing the upper half plane. Analogously to the way in which we found III(1.3), we also find

$$iG(k,t)\,|_{t=-0} = \int_0^\infty A_-(k,\omega)\,d\omega = \frac{i}{2\pi}\int_C G(k,\omega)\,d\omega \qquad \text{III}(2.1)$$

Then, from eq. II(3.1),

$$< 0\,|T|\,0 > = i \int_C \frac{d^4k}{(2\pi)^4}\,\frac{k^2}{2M}\,G(k,\omega) \qquad \text{III}(3)$$

where we have changed the sum in eq. II(3.1) into an integral. Combining (2) and (3), we find

$$E_{int} = (0\,|\,V\,|0) = \frac{i}{2}\int_C \frac{d^4k}{(2\pi)^4}\left(\omega - \frac{k^2}{2m}\right)G(k,\omega) \qquad \text{III}(4)$$

where C indicates the contour on the ω integration.

Now our general expression for $G(k,\omega)$ is

$$G(k,\omega) = \frac{1}{\dfrac{k^2}{2m} - \omega + \Sigma(k,\,\omega) - i\,\delta\,\mathrm{sgn}\,\omega}\,.$$

Since $\omega G(k,\omega) \rightarrow 1$ as $\omega \rightarrow \infty$, the above formula (4) isn't practical but we can calculate the difference in interaction energy H_{int} between the interacting and noninteracting systems. This can be written

$$\frac{1}{2i}\frac{1}{(2\pi)^4}\int_C \left\{ \frac{k^2/2m - \omega + \Sigma(k,\omega)}{\frac{k^2}{2m} - \omega + \Sigma(k,\omega) - i\,\delta\,\mathrm{sgn}\,\omega} - \frac{k^2/2m - \omega}{\frac{k^2}{2m} - \omega - i\delta\,\mathrm{sgn}\,\omega} \right\} d^4k$$

$$- \frac{1}{2i}\frac{1}{(2\pi)^4}\int_C \Sigma(k,\omega)\, G(k,\omega)\, d^4k \qquad \text{III}(4.1)$$

and since the first two terms cancel, we find

$$E_{int} = -\frac{1}{2i}\int_C \frac{d^4k}{(2\pi)^4}\quad \Sigma(k,\omega)\, G(k,\omega) \qquad \text{III}(5)$$

and the total energy

$$E_0 = i\int \frac{d^4k}{(2\pi)^4}\left[\frac{k^2}{2m} + \frac{1}{2}\Sigma(k,\omega)\right] G(k,\omega) \qquad \text{III}(6)$$

We next obtain an expression for the total energy E in terms of the interaction energy alone. Let us work with H in the form

$$H(g) = T + g\,V \qquad \text{III}(7)$$

where we obtain the H of the physical system we are dealing with for $g = 1$. Now

$$E_0(g) = \langle 0 | T | 0 \rangle + \langle 0 | gV | 0 \rangle \qquad \text{III}(8)$$

and

$$\frac{\partial E_0}{\partial g} = \langle 0 | V | 0 \rangle \qquad \text{III}(8.1)$$

even though the wave functions $| 0) = \Psi_0$ are functions of g, because of the minum property of the expectation value of the Hamiltonian taken with the physical wave functions. Thus,

$$E_0 = \int_0^g \frac{dg}{g} E_{int.}(g) + \text{Constant}$$

$$= \int_0^g \frac{dg}{g} E_{int}(g) + (\Phi_0 | T | \Phi), \qquad \text{III}(9)$$

where

$$E_{int}(g) = (\delta_0(g) | gV | \delta_0(g)) \qquad \text{III}(9.1)$$

is the interaction energy for coupling constant g, δ_0 (g) being the physical ground-state wave functions for the relevant g. The constant in III(9) is $(\varphi_0 | T | \varphi_0)$, the expectation value of T in the bare ground state φ_0: (to which δ_0 tends as $g \to 0$).

Eq. (9) gives us a way of connecting the Goldstone-Hugenholtz expansion for E_0 discussed in part I with the Green's Function expression III(4). In the Goldstone expansion, the corrections to E_0 are all expressed as corrections to the potential energy, in the form III(9), the kinetic energy term $(\varphi_0 | T | \varphi_0)$ being just the kinetic energy of the bare ground state as in the Goldstone expression. The integral over g brings in a factor 1/n in the nth order term in $E_{int}(g)$. Thus, $E_{int}(g)$ must be just n times the nth order term in the Goldstone expansion, $E^{(n)}_{Goldstone}$; i.e.,

$$E^{(n)}_{int}(g) = n\, E^{(n)}_{Goldstone} \quad . \qquad \text{III(10)}$$

Then we must also have

$$T^{(n)} = -(n-1)\, E^{(n)}_{Goldstone} \,, \; n > 1 \qquad \text{III(10.1)}$$

where $T^{(n)}$ is the nth order term in the kinetic energy $(\Phi_0 | T | \Phi_0)$, in order to give the same final result.

It is instructive to work through this connection for the second-order term in perturbation theory. The first-order terms are too trivial to be instructive, and we shall drop them. We use Green's functions without self-energy insertions, since these come in higher order*.

A simple term of 2nd order in g in $E_{int}(g)$ is shown in Fig. 16. It is made by joining the self-energy insertion $\Sigma(k, \omega)$ of Fig. 14 together with $G(k,\omega)$ as indicated in eq. III(3). This differs from the corresponding Goldstone diagram by the fact that the lines

*One great advantage of the Green's function expression III(4) is that it is in terms of "clothed" propagators which involve physical occupation numbers, etc. It is difficult to express the Goldstone expansion in terms of physical occupation numbers.

do not have arrows; i.e., $G(k,\omega)$ can refer to either particle or hole propagation.

Now $\Sigma(k,E)$ is given by eq. II(20), so that

$$E_{int} = \frac{i}{2}\int \frac{d^4q}{(2\pi)^4}\frac{d^4k}{(2\pi)^4} V_q^2 G(k,E)G(k-q, E-\omega)\,\pi_0(q,\omega) \qquad \text{II(11)}$$

$$= \frac{i}{2}\int \frac{d^4q}{(2\pi)^4} V_q^2 \pi_0^2(q,\omega).$$

Using eqs. II(20.2), (20.3) (suppressing the factor 2 from the spin sum) we find this equal to

$$E_{int} = \frac{i}{2}\int \frac{d^4q}{(2\pi)^4} V_q^2 \left\{ \int_{\substack{p>k_F\\ |p-q|<k_F}} \frac{d^3p}{(2\pi)^3}\frac{1}{T_p - T_{p-q} - \omega - i\delta} - \int_{\substack{p<k_F\\ |p-q|>k_F}} \frac{d^3p}{(2\pi)^3}\frac{1}{T_p - T_{p-q} - \omega + i\delta}\right\}^2$$

where we have kept only the kinetic energies in the denominator, since self-energy corrections are of higher order. Because of the position of the poles,

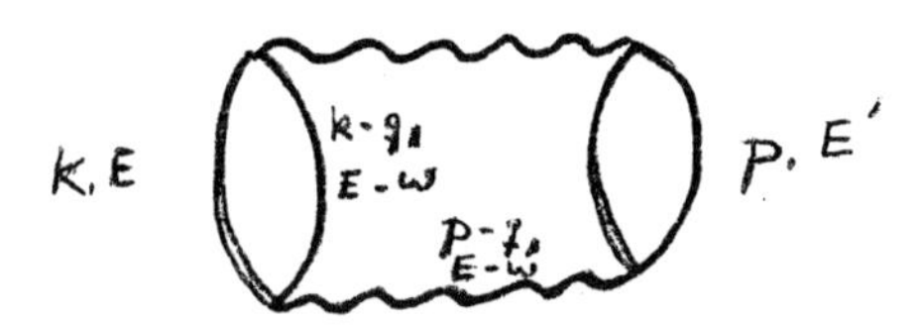

Fig. 16. Second-order diagram for $E_{int}(g)$ in terms of Green's Functions.

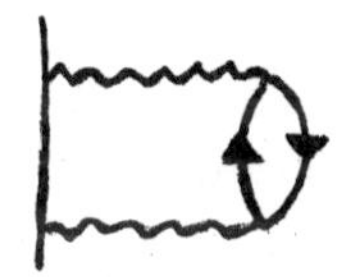

Fig. 17. Self-energy part which would be obtained by differentiating Fig. 16.

only the two cross terms contribute in the square, and we find

$$E_{int} = - \int \frac{d^3q}{(2\pi)^3} V_q^2 \int \frac{d^3p}{(2\pi)^3} \int \frac{d^3k}{(2\pi)^3} \frac{1}{T_p + T_{k-q} - T_k - T_{p-q}}$$

$$\begin{array}{ll} p > k_F & k > k_F \\ |p - q| < k_F & |k - q| > k_F \end{array}$$

which is just 2 times the $W^{(2)}$ of the Goldstone expansion, given in eq. I(22).

According to our general argument, the 2nd order correction to the kinetic energy must simply cancel half of this potential energy.

The single-particle energy is defined as

$$\frac{\delta E_0}{\delta n(k_1)} = \epsilon(k_1). \qquad \text{III(12.1)}$$

To see how things go, we assume the only nonzero self-energy insertion to be as shown in Fig. 17.

The functional variation here is with respect to quasiparticle number. We add one bare particle to the bare ground state; this simply changes our classification of state k_1 from "empty" to "full". Various effects of this will manifest themselves in the "clothed", or physical, Green's Function $G(k,\omega)$, being propagated in various ways through the interaction. Thus, an assumption is involved here that bare particles are transformed into quasiparticles by the interactions*.

The point here is that the classifications of states as particle states or hole in the clothed propagator $G(k, \omega)$ is in terms of the <u>bare</u> ground state, so that we can vary with respect to quasiparticle number $n(k_1)$ by varying the bare ground state with respect to $n_0(k_1)$, the occupation number in the bare ground state, and then calculating how this change propagates.

Since, in the neighbourhood of the quasiparticle pole,

$$G(k,\omega) = \frac{z_k}{T_k + \Sigma(k, \epsilon_k) - \epsilon_k - i\delta}$$

*See <u>Interacting Fermi Systems</u>, P. Nozières, W. A. Benjamin, Inc., New York, 1964. The so called "Adiabatic Hypothesis", p. 168 and Ch. 6, § 76, p. 291 is especially relevant to the discussion here.

for particles, and

$$G(k, \omega) = \frac{Z_k}{T_k + \Sigma(k, \epsilon_k) - \epsilon_k + i\delta}$$

for holes,

$$\frac{\delta G(k, \omega)}{\delta n(k_1)} = Z_{k_1} \delta(\underline{k} - \underline{k}_1) \delta(\omega - \epsilon_{k_1}) \qquad \text{III}(13)$$

$$- G^2(k) \frac{\delta}{\delta n(k_1)} \left[T_k + \Sigma(k, \omega)\right]$$

$$= \frac{-G^2(k) \delta \Sigma(k, \omega)}{\delta n(k_1)}$$

The second term on the right-hand side comes from variation of the part of $G(k, \omega)$ other than that relating to the quasiparticle pole. Now T_k does not depend upon $n(k_1)$, so that only the $\delta \Sigma(k, \omega)/\delta n(k_1)$ enters in that term.

Rather than carry out the functional differentiation indicated in eq. (12.1) we shall show in detail how the second functional differentiation is carried out; namely, we find $\delta^2 E/\delta n(1)\, \delta n(2)$.

First we note that

$$\frac{\delta \epsilon_1}{\delta n(2)} = \delta_2 \epsilon (k_1, \epsilon_1) + \frac{\partial \Sigma}{\partial \omega} \Big|_{\omega = \epsilon_1} \frac{\delta \epsilon_1}{\delta n(2)}$$

III(14)

$$= Z_1 \delta_2 \epsilon (k_1, \epsilon_1) = z_1 \delta_2 \Sigma (k_1, \epsilon_1)$$

where $\delta_2 \epsilon(k_1, \epsilon_1)$ means the variation with respect to addition or removal of a quasiparticle of momentum k_2 without change of quasiparticle energy ϵ_1 occurring in internal lines in the graphs for Σ. Thus, δ_2 is a "partial" functional derivative. Of course, addition or removal of quasiparticle 2 will change ϵ_1, which runs through the whole diagram for $\Sigma(k_1, \epsilon_1)$, and this change is taken into account by the factor Z_1. There is no variation in the kinetic energy $k_1{}^2/2m$, which is independent of n(2).

We write $\Sigma(k_1, \omega_1)$ as

$$\Sigma (k_1, \omega_1) = +i \int \frac{d^4 k_2}{(2\pi)^4} v(k_2, \omega_1; k_2 \omega_2) \, G(k_2, \omega_2)$$

III(15)

separating off the $G(k_2, \omega_2)$ we wish to differentiate in all possible ways, which defines a particle-hole interaction v. We find then

$$\delta_2 \Sigma(k_1, \omega_1) = + Z_2 \int v(k_1,\omega_1;k,\omega) G(k,\omega) \, \delta(\underline{k} - \underline{k}_2) \, \delta(\omega - \epsilon_2) \frac{d^4k}{(2\pi)^3}$$

$$-i \int v(k_1,\omega_1; k,\omega) \, G^2(k,\omega) \frac{\delta \Sigma(k,\omega)}{\delta n(2)} \frac{d^4k}{(2\pi)^4}$$

III(15.1)

and

$$\frac{\delta}{\delta n(2)} \epsilon_1 = z_1 \delta_2 \Sigma(k_1, \epsilon_1) = + z_1 z_2 v(k_1, \epsilon_1; k_2, \epsilon_2)$$

$$- iZ_1 \int v(k_1, \epsilon_1; k, \omega) \, G^2(k,\omega) \frac{\delta \Sigma(k,\omega)}{n(2)} \frac{d^4k}{(2\pi)^4} .$$

III(16)

One can think of iterating this equation, taking $-Z_2 v$ as the first approximation for $\delta_2(k_1, \epsilon_1)$. Such iteration produces those particle-hole graphs as obtained by breaking open self-energy insertions, as shown in Fig. 32. The above equation is equivalent to Landau's equation for Γ^ω, $\Gamma^{(1)}$ being obtained from v by putting momenta on the Fermi surface.

QUANTUM ELECTRODYNAMICS AT HIGH ENERGY

Laurie M. Brown

QUANTUM ELECTRODYNAMICS AT HIGH ENERGY

Laurie M. Brown
Northwestern University

LECTURE I

1. - Introduction

Recently there has been a revival of interest in calculations in quantum electrodynamics, due mainly to the increase in accuracy of experiments at low energy (Lamb Shift, Magnetic moment, hyperfine structure) and to the possibility of performing very high energy experiments with new accelerators and storage rings. Both low energy and high energy experiments require the calculation of higher order corrections in perturbation theory: low energy because of the high precision sought, and high energy because these corrections are not small. Higher order corrections are generally complicated and cumbersome in form, lengthy to calculate, and not very transparent from the physical point of view. On the other hand, the high energy limits are usually much simpler. Thus there would be considerable advantage in being able to write down the high energy limits directly.

Before proceeding, let me make a remark about why we should be interested in very high energy experiments in quantum electrodynamics. The obvious answer is that we wish to test the validity of Q.E.D., the only field theory we have which seems to be in reasonably good repute. So far no experiments at high energy have turned up any apparent violations of the theory,[1] but at some level of energy and/or accuracy they are bound to - because the effects of strongly interacting particles in intermediate states must show up. For example, in addition to vacuum polarization loops of electrons and muons there will also be loops of pions. This effect is sensitive to the pion form factor in the timelike region of momentum transfer, which is related to the formation of the rho meson. So at a certain point we shall be seeing this and other strong interaction effects, especially in electron-positron annihilation, and to interpret them we must have a very good idea of the pure Q.E.D. part.

In general there are two kinds of labor involved in calculating higher order processes: tedious algebraic manipulations and evaluating integrals. Certain methods have been developed to pick out leading terms in evaluating the integrals,[2] and what I will talk about mainly is simplifying the algebra.

The theory will be formulated in terms of two-component helicity spinors describing fermions of non-zero mass. This description of electrons was developed earlier, but is not very well known, and its main features will be described in the first lecture.[3] The perturbation expansion of its S-matrix contains propagators of Klein-Gordon type and its vertex interaction is the Klein-Gordon interaction plus a spin-dependent term. In particular, it contains the Klein-Gordon "double corner" terms, corresponding to the simultaneous interaction of two photons with the electron.

In the second lecture we shall show that these terms can always be eliminated, leaving a simpler expression (Rule I). At high energies this expression can be expanded in powers of m/E where m and E are mass and energy. The leading term of this expansion is particularly simple, and we will further show that it approximately conserves fermion helicity along the Feynman line (cohelicity: i.e., helicity for fermions and negative helicity for antifermions), a fact which leads to further simplification (Rule II).[4]

In the third lecture we shall illustrate by several examples the calculation of high energy limits.

2. - The Two-Component Electron Theory

The equations

$$
\begin{aligned}
i\partial^{-}\psi(x) &= m\,\Omega(x) \\
i\partial^{+}\Omega(x) &= m\,\psi(x)
\end{aligned}
\tag{1}
$$

where $\partial^{\pm} = \partial_0 \mp \underline{\sigma} \cdot \nabla$ and $\psi(x)$ is a two-component spinor satisfying the free Klein-Gordon equation, give a factorization of that equation as can be seen by substitution:

$$(\Box + m^2)\,\psi(x) = 0, \tag{2}$$

and also

$$(\Box + m^2)\;\Omega(x) = 0. \tag{3}$$

Equations (1), therefore, are a representation of Dirac's equation for a four-component spinor $\chi(x) = (\Omega(x), \psi(x))$ and can, in fact, be written as

$$(i\partial_0 \quad i\,\underline{\alpha}\cdot\nabla)\,\chi(x) = \beta\, m\, \chi(x) \tag{4}$$

with

$$\underline{\alpha} = \begin{pmatrix} -\underline{\sigma} & 0 \\ 0 & \underline{\sigma} \end{pmatrix}, \qquad \beta = \begin{pmatrix} 0 & I \\ I & 0 \end{pmatrix}. \tag{5}$$

For momentum eigenfunctions, in the limit $m \to 0$, we get

$$p^{-}\psi = p^{+}\Omega = 0. \tag{6}$$

In general, we use the notation

$$a^{\pm} = a_0 \pm \underline{\sigma}\cdot\underline{a} \tag{7}$$

where a_μ is a four-vector (note the special convention for $\partial^{\pm}$). With $\underline{\nu} = \underline{p}/|\underline{p}|$, and p_0 assumed positive, Eqs. (6) can be written

$$\begin{aligned} \underline{\sigma}\cdot\underline{\nu}\,\psi &= \psi \\ \underline{\sigma}\cdot\underline{\nu}\,\Omega &= -\,\Omega \end{aligned} \tag{8}$$

Showing that ψ and Ω are respectively positive and negative helicity eigenstates. They are just the equations of the two-component neutrino theory.

Our present interest is not in the special case $m = 0$, but rather

$$m \ll E = (\underline{p}^2 + m^2)^{1/2},$$

as we seek an expansion in powers of m/E. However, the two-component electron theory is convenient to use in discussing the high energy limit just because the momentum states are helicity eigenstates in that limit.

By using gauge invariance in the usual way we obtain the generalization of Eqs. (1) in the presence of an electromagnetic field, i.e. replace $i\partial_\mu$ by $i\partial_\mu - eA_\mu = \pi_\mu$, obtaining

$$\left.\begin{aligned} \pi^- \psi(x) &= m\,\Omega(x), \\ \pi^+ \Omega(x) &= m\,\psi(x), \end{aligned}\right\} \quad (9)$$

and the corresponding second order equation obeyed by $\psi(x)$:

$$\pi^+\pi^- \psi(x) = m^2\psi(x), \quad (10)$$

which we call the Kramers Equation (sometimes called the equation of Feynman and Gell-Mann). This eq. (10) for the two-component spinor $\psi(x)$ is taken as the basic equation of our theory, hence the name two-component fermion theory. The spinor $\Omega(x)$ is merely a derivative of $\psi(x)$ (via Eq. (9)) and provides no new information.

Let me make some brief remarks concerning some interesting features of this equation, and then discuss one or two in more detail.

a. - Applied to pure electrodynamics the results are identical to those obtained using the Dirac equation, although the formalism is rather different. Like the Dirac equation, we can prove the Lorentz covariance of the Kramers equation, but ψ and Ω have different relativistic transformation properties. While the Dirac Equation is parity (P) and charge conjugation (C) invariant, the Kramers Equation is not, but it is CP invariant. The antiparticle state is identified as the negative frequency state ($-p_0 = E$ for a free particle) and is interpreted as the CP-conjugate, not the C-conjugate state. Thus for a free particle state

$$u\, e^{-ip \cdot x} = u\, e^{-i(p_0 t - \mathbf{p} \cdot \mathbf{x})}$$

with

$$p_0 = - E = - (\underline{p}^2 + m^2)^{1/2},$$

the momentum is $\mathbf{p}$ and <u>not</u> $-\mathbf{p}$.

b. - It is not possible to give the Kramer's particle a phenomenological anomalous magnetic moment, except in a static potential, without the theory becoming inconsistent. That is, one cannot include a term analogous to the Pauli magnetic moment. In this sense the theory implies the principle of minimal electromagnetic interaction, that all moments are due to currents.

c. - The C and P violating, but CP conserving, property of the Kramers Equation suggested to Feynman and Gell-Mann the V-A theory of weak interactions.

d. - The Kramers Equation can be derived from a Lagrangian density

$$\mathcal{L}(x) = m^{-1}(\pi^{+}\Omega)^{+}(\pi^{-}\psi) - m\,\Omega^{+}\psi. \tag{11}$$

From this Lagrangian the entire field theory and S-matrix expansion can be obtained by standard methods (Tonin[3]). The results, except for the use of anticommutation in place of commutation resemble the Klein-Gordon (or, more precisely, the Pauli-Weisskopf) results, surprisingly, more than the Dirac results. For example, the anticommutation rules are:

$$[\psi(x),\ \Omega^{+}(y)]_{+} = i\,m\,\Delta(x-y), \tag{12}$$

all other anticommutators of fields being zero.

The most important result we shall need from the canonical formalism is the current density:

$$s_{\mu}(x) = \psi^{+}(x)\sigma_{\mu}\psi(x) + \Omega^{+}(x)\,\overline{\sigma}_{\mu}\,\Omega(x), \tag{13}$$

with $\sigma_{\mu} = (1, \underline{\sigma})$ and $\overline{\sigma}_{\mu} = (1, -\underline{\sigma})$. Applying the divergence operator $\partial_{\mu} \equiv (\partial_{0}, -\underline{\nabla})$ and using Eqs. (1) or (9), we see that this current is conserved. For plane wave states

$$\psi(x) = \psi\, e^{-ip\cdot x} \tag{14}$$

with ψ a constant spinor, from Eq. (1)

$$\Omega(x) = \frac{1}{m}\, p^{-}\,\psi(x)$$

$$= \Omega\, e^{-ip\cdot x} \tag{15}$$

with

$$p^- \psi = m\,\Omega$$

$$p^+ \Omega = m\,\psi. \tag{16}$$

The charge density becomes

$$s_0 = \psi^+ \psi + \Omega^+ \Omega$$

$$= m^{-2}\, \psi^+ \,(m^2 + p^- p^-)\, \psi. \tag{17}$$

Using

$$(\underline{\sigma} \cdot \underline{a})\,(\underline{\sigma} \cdot \underline{b}) = \underline{a} \cdot \underline{b} + i\,\underline{\sigma} \cdot \underline{a} \times \underline{b} \tag{18}$$

we have

$$p^- p^- = p_0^2 - 2p_0\,(\underline{\sigma} \cdot \underline{p}) + p^2$$

and

$$p^- p^- + m^2 = 2p_0 (p_0 - \underline{\sigma} \cdot \underline{p}) = 2p_0 p^-,$$

so that

$$s_0 = \frac{2p_0}{m^2}\; \psi^+ p^- \psi$$

$$= \frac{2p_0}{m}\; \psi^+ \Omega$$

$$= \frac{2p_0}{m}\; \Omega^+ \psi. \tag{19}$$

We observe that Eq. (17) implies that s_0 is positive, for either sign of p_0. Therefore $\Omega^+\psi$ must itself have the sign of p_0. Defining $\epsilon = p_0/|p_0|$ we choose the normalization

$$\Omega^+\psi = \psi^+\Omega = \epsilon , \tag{20}$$

so that the particle density is given by

$$s_0 = \frac{2p_0\epsilon}{m} = \frac{2E}{m} . \tag{21}$$

An important result of this choice of normalization is the following:

We define the scalar product of two states in four-dimensional form so that for plane waves of momentum $\underline{p}$ and q we get

$$< \Omega_p | \psi_q > = \epsilon_p \delta^4 (p - q) \tag{22}$$

As a consequence the projection of a scattering state ψ_i (x) on a plane wave state ψ_p(x) is

$$< \Omega_p | \psi_i > \tag{23}$$

In other words, S-matrix elements are taken between a ψ on the right and an Ω on the left.

This is also necessary from the standpoint of the Lorentz transformation properties of ψ and Ω. A brief excursion will show why this is so:

Consider a three-vector $\mathbf{p}$, which we can represent by the 2x2 matrix $\boldsymbol{\sigma} \cdot \mathbf{p} \equiv \hat{p}$, since $p_i = T_r[\sigma_i \hat{p}]$. Choosing an arbitrary direction which we represent by the <u>matrix</u> $\hat{u} = \boldsymbol{\sigma} \cdot \mathbf{u}$, where $\mathbf{u}$ is a unit vector, we resolve $\mathbf{p}$ or $\hat{p}$ into components parallel and perpendicular to $\hat{u}$,

$$\hat{p} = \hat{p}_{\parallel} + \hat{p}_{\perp} \tag{24}$$

We can then easily show that $\hat{p}_{\parallel}$ commutes and $\hat{p}_{\perp}$ anticommutes with $\hat{u}$. Now rotate the coordinate system about $\hat{u}$ through the angle α; then the components of $\mathbf{p}$ become $p_i' = \mathrm{Tr}\,[\sigma_i \hat{p}']$, where

$$\hat{p}' = R\,\hat{p}R^{-1}, \qquad R = e^{i\frac{\alpha}{2}\hat{u}} \ . \tag{25}$$

Proof:

$$R\hat{p}R^{-1} = R(\hat{p}_{\parallel} + \hat{p}_{\perp})R^{-1}$$

$$= \hat{p}_{\parallel} + R^2 \hat{p}_{\perp}, \tag{26}$$

because of the commutation properties of $\hat{p}_{\parallel}$ and $\hat{p}_{\perp}$ with $\hat{u}$. Since $R^2 = e^{i\alpha\hat{u}} = \cos\alpha + i\hat{u}\sin\alpha$, using Eq. (18) we obtain

$$R^2 \hat{p}_{\perp} = \hat{p}_{\perp} \cos\alpha + \boldsymbol{\sigma} \cdot (\mathbf{p}_{\perp} \times \mathbf{u}) \sin\alpha \tag{27}$$

To see what this means, let $\hat{u} = \sigma_z$ and $\hat{p}_{\perp} = p_{\perp}\, \sigma_y$;

$$R^2 \hat{p}_{\perp} = p_{\perp} (\sigma_y \cos\alpha + \sigma_x \sin\alpha),$$

and

$$p_y' = p_{\perp} \cos\alpha$$

$$p_x' = p_{\perp} \sin\alpha$$

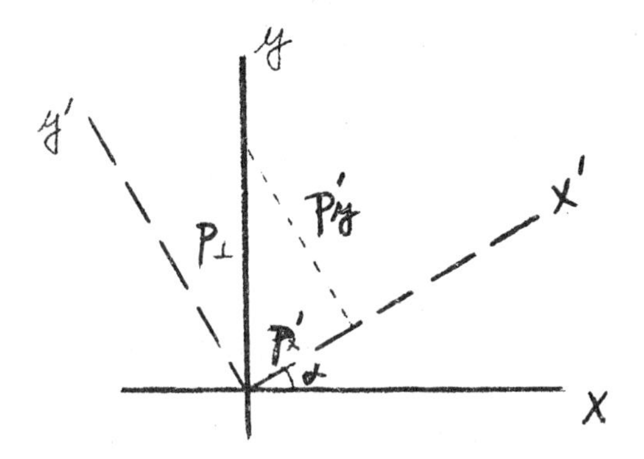

Fig. 1.

It is clear that under a rotation of coordinates, also

$$(\hat{a}\hat{b}\,\hat{c}\ldots)' = R(\hat{a}\hat{b}\,\hat{c}\ldots)R^{-1}, \tag{28}$$

and that

$$a'^{+} = Ra^{+}R^{-1},\quad a'^{-} = R\,a^{-}R^{-1}, \tag{29}$$

as well as

$$(a^{+}b^{-}c^{-}\ldots)' = R(a^{+}b^{-}c^{-}\ldots)R^{-1}. \tag{30}$$

In particular this applies to the operator $\pi^{+}\pi^{-}$ in the Kramers equation for ψ (and to $\pi^{-}\pi^{+}$ in the equation for Ω) so that

$$\psi' = R\,\psi,\quad \Omega' = R\,\Omega \tag{31}$$

for the equations to be rotation invariant.

Similarly, for a pure Lorentz transformation with velocity $\beta\mathbf{u}$ along the direction $\mathbf{u}$ we get

$$a'^{+} = l\, a^{+}\, l, \quad a'^{-} = l^{-1}\, a^{-}\, l^{-1} \tag{32}$$

where

$$l = e^{-\frac{\alpha}{2}\hat{u}}, \quad l^{-1} = e^{\frac{\alpha}{2}\hat{u}}, \quad \text{and } \alpha = \tanh^{-1}\beta. \tag{33}$$

Exactly the same method is used to show this (see Boulder lectures[3]). From this we conclude that

$$(\pi^{+}\pi^{-})' = l\,\pi^{+} l \cdot l^{-1} \pi^{-} l^{-1} = l(\pi^{+}\pi^{-})\, l^{-1}, \tag{34}$$

so that

$$\psi' = l\,\psi. \tag{35}$$

On the other hand,

$$(\pi^{-}\pi^{+})' = l^{-1}\,(\pi^{-}\pi^{+})\; l, \tag{36}$$

so that

$$\Omega' = l^{-1}\,\Omega \tag{37}$$

Thus, while $< \Omega_i \mid \psi_j >$ is Lorentz invariant, $< \psi_i \mid \psi_j >$ is not.

LECTURE II

(The work presented in this lecture is based on reference 4).

I. Feynman Rules

We begin by writing the Kramers Equation in the form

$$(P^+ - eA^+(x))(P^- - eA^-(x))\psi(x) = m^2\psi(x) \quad (38)$$

where

$$P^\pm \equiv i\partial^\pm = i(\partial_0 \mp \underline{\sigma}\cdot\nabla).$$

Note that $\partial^+\partial^- = \Box$, and that for any two commuting four-vectors a_μ, b_μ

$$a^+b^- + b^+a^- = 2\,a_\mu b_\mu. \quad (39)$$

We can, therefore, write Eq. (38) as

$$(\Box + m^2)\,\psi(x) = -\,J(x)\,\psi(x) \quad (40)$$

with

$$J(x) = e\,(P^+A^-(x) + A^+(x)\,P^-) - e^2A^+(x)\,A^-(x). \quad (41)$$

From this we see that the virtual electron propagator in momentum space is just $(p^2 - m^2)^{-1}$, as in the Klein-Gordon case, and that there are vertex operators for single and double corners as follows:

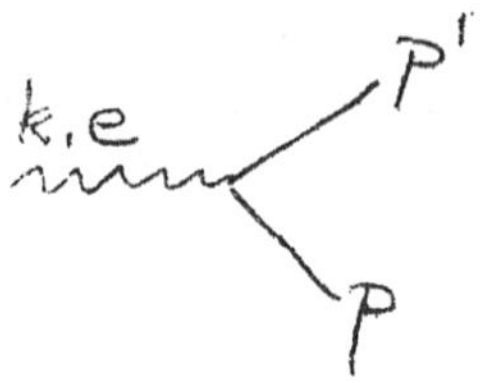

Kramers $\quad p'^{+} e^{-} + e^{+} p^{-}$

(Klein-Gordon: $\quad p' \cdot e + e \cdot p$) $\quad$ (42)

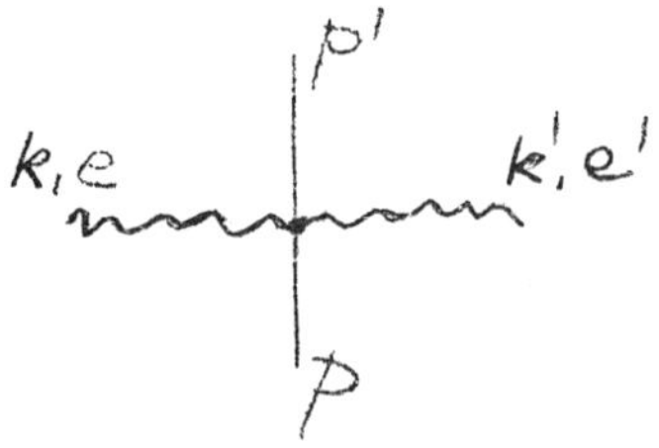

Kramers: $\quad -e^{+} e'^{-}$

(Klein-Gordon: $\quad -e \cdot e'$) $\quad$ (43)

Note that permuting the two photons gives rise to another double corner, so that in the Klein-Gordon case we have, effectively, $-2e \cdot e'$. Also in the Kramers case we get

$$- e'^{+} e^{-} - e^{+} e'^{-} = -2e \cdot e' \quad (44)$$

The complete set of Feynman rules in momentum space has been given by Tonin.[3] They are exactly the same as the rules for the Klein-Gordon theory as given in many standard texts[5] providing we use the vertex interactions given in Eqs. (42) and (43), multiply each graph by a factor $(-)^{l}\delta_{\underline{p}}$, where l is the number of closed fermion loops and $\delta_{\underline{p}}$ is the sign of permutation of the final fermions, and provided we use two-component spinors normalized according to Eq. (20). In calculating cross sections, however, we must recall Eq. (21) which says the particle density is 2E/m and <u>not</u> E/m.

For graphs containing only single corners, alternatively, the rules are just the standard ones for the Dirac theory with Klein-Gordon propagators, Eq. (42) replacing $\not{e}$, and use of two-component spinors. We shall see, in section IV, that indeed only such graphs (slightly modified) need be considered.

II. Helicity Spinors

We now specify the two-component helicity spinors for the case. First introduce a set of two-component spinors

$$\chi_s, \quad s = \pm 1,$$

obeying

$$\boldsymbol{\sigma} \cdot \mathbf{p}\, \chi_s = s\, |\mathbf{p}|\, \chi_s \tag{45}$$

and normalized

$$\chi_s \chi_{s'} = \delta_{ss'} \tag{46}$$

These specify the actual helicity, since the three-vector $\mathbf{p}$ appearing in the wave functions $\psi_{\mathbf{p}}(x)$ and

$\Omega_{\mathbf{p}}(x)$ is the actual momentum and not $\epsilon\,\mathbf{p}$. This is because particle-antiparticle conjugation in this theory is CP and not C.

Next consider a plane wave solution of Eqs. (1) of the form (with f(x) a normalized plane wave):

$$\psi_s(x) = a(s,\epsilon)\,\chi_s f(x) \tag{47}$$

so that

$$\Omega_s(x) = m^{-1}(p_0 - s\,|\mathbf{p}|)\,a(s,\epsilon)\,\chi_s\,f(x). \tag{48}$$

Then, the normalization condition (20) gives

$$\epsilon = \Omega_s^+ \psi_s = m^{-1}(p_0 - s\,|\mathbf{p}|)\,|a(s,\epsilon)|^2, \tag{49}$$

and hence

$$|a(s,\epsilon)|^2 = \frac{m}{E - \epsilon s\,|\mathbf{p}|} = \frac{E + \epsilon s\,|\mathbf{p}|}{m}.$$

In terms of the cohelicity, $S \equiv \epsilon s$, we then write

$$a_S \equiv a(s,\epsilon) = \left(\frac{E + S\,|\mathbf{p}|}{m}\right)^{1/2} \tag{50}$$

and finally,

$$\left.\begin{aligned} \psi_s &= a_S\,\chi_s \\ \Omega_s &= \epsilon\,a_{-S}\,\chi_s \end{aligned}\right\} \tag{51}$$

Matrix elements of any matrix M are given by

$$M_{TS} = (\Omega\ (p', t),\ M\psi(p, s)) = \epsilon\ ' a'_{-T} a_S \, .$$

$$< \chi_t \mid M \mid \chi_s > . \tag{52}$$

We note that $a_+ a_- = 1$, and that in the high energy limit

$$a_+ \to \left(\frac{2E}{m}\right)^{1/2} \quad \text{while} \quad a_- \to \left(\frac{m}{2E}\right)^{1/2} .$$

III. Elimination of Double Corners

Consider a part of a Feynman graph consisting of a single fermion line and its consecutively numbered interactions, n in number. We call this part of the graph M_n.

M_n consists of N(n) contributions, including one term containing only single corners, the remaining terms containing one or more double corners. By considering the two possibilities (single or double corner) for the nth interaction, say, we see that N obeys the recursion relation

$$N(n) = N(n - 1) + N(n - 2), \tag{53}$$

and grows rapidly with n.

However, let us focus our attention on the rth vertex and consider the N(n) contributions to M_n grouped according to whether the rth vertex is single, is doubled with the preceding vertex, or is doubled with the succeeding vertex. Let us also, for typographic simplicity, shift the numeration of the graph parts so r = 0. We are then considering the sum of the parts of M_n diagrammed below

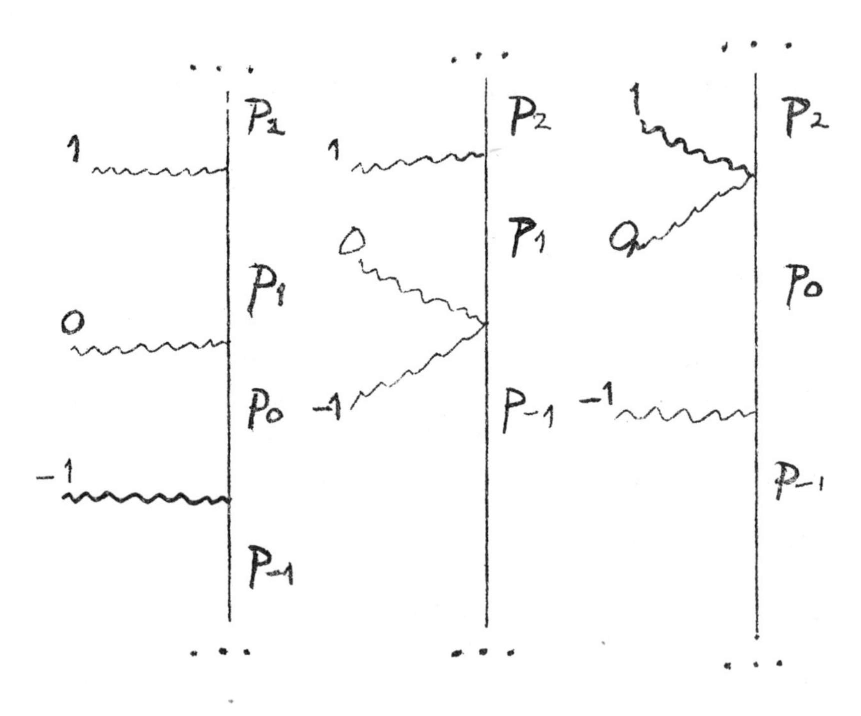

Fig. 2.

Letting the single corner interaction, Eq. (21), for the rth vertex be denoted by V_r, we have

$$M_n = \sum \ldots [V_1 V_0 V_{-1} - V_1 e_0^+ e_{-1} (p_0^2 - m^2) - e_1^+ e_0^- V_{-1} (p_1^2 - m^2)] \ldots \tag{54}$$

where ... represents common factors and Σ means we are to sum over all possible arrangements of the interactions, other than the vertex labelled zero, into single and double corners, without changing the ordering. In multiplying out $V_1V_0V_{-1}$ two terms containing p_0^2 and two terms containing p_1^2 appear, since

$$p^+p^- = p^2 ;$$

these are respectively

$$V_1e_0^+e_{-1}^- p_0^2 \qquad \text{and} \qquad e_1^+e_0^-V_{-1} p_1^2 .$$

Therefore the bracketed expression in Eq. (34) is just $V_1V_0V_{-1}$ with p_0^2 and $p_1{}^2$ each replaced by m^2. We thus arrive at

Rule I: The matrix corresponding to the N(n) terms of a fermion line with n interactions is obtained from the matrix containing only single corners by replacement of $p_r{}^2$ by m^2 (not, of course, in the propagators).

IV. High Energy Limit

We consider now matrix elements of M_n, Eq. (54) taken between helicity states as in Eq. (52). According to Rule I M_n, aside from propagators, is given by

$$\prod_{i=n}^{1} V_i = V_n \dots V_1, \tag{55}$$

in which $p_i{}^2$ is replaced by m^2. We write

$$V_i = A_i + B_i \tag{56}$$

with

$$A_i = p^+_{i+1}\, e^-_i, \quad B_i = e^+_i\, p^-_i\,. \tag{57}$$

Then in writing out Eq. (35), a factor m^2 will result whenever B_i is followed by A_{i-1}. In the relativistic limit we drop these terms.

The remaining terms are then

$$\prod_{i=n}^{1} A_i, \quad \prod_{i=n}^{1} B_i, \text{ and } \prod_{j=n}^{l} A_j \prod_{i=l-1}^{1} B_i, \quad l = 2 \ldots n \tag{58}$$

Using

$$p^{\pm}\, \chi_s = \epsilon\, m\, a_{\pm S}\, \chi_s, \tag{59}$$

we denote the contributions of terms (58) to M_{TS}, as given by Eq. (52) with the notation:

$$< \chi_t(p'),\ M\, \chi_s\,(p) > = M_{ts}, \tag{60}$$

and have

$$F'_{TS} = \epsilon'\, a'_{-T}\, a_S \left(\prod_n^1 A_i\right)_{ts} = m a'_T a_S \left(e_n \prod_{n-1}^{1} A_i\right)_{ts}, \tag{61}$$

$$F''_{TS} = \epsilon'\, a'_{-T} a_S \left(\prod_n^1 B_i\right)_{ts} = m\, a'_{-T}\, a_{-S}\, \epsilon\epsilon' \left(\prod_n^2 B_i e_1^+\right)_{ts}, \tag{62}$$

$$F'''_{TS} = \epsilon' a'_{-T} a_{S'} {}_t\left(\prod_{j=n}^{1} A_j \prod_{l-1}^{1} B_i \right)_s$$

$$= \epsilon\, m^2 a'_T a_{-S} {}_t\left(e_n^- \prod_{j=n-1}^{1} A_j \prod_{l-1}^{2} B_i e_1^+ \right)_s, \tag{63}$$

$$l = 2 \ldots n.$$

Next we investigate how large these expressions can become. For this purpose we approximate each internal momentum by its energy. If all energies are large, then when F'''_{TS} is largest, i.e. when $T = -S = 1$, it is of order m/E relative to F'_{TS} and F''_{TS} when they are largest (i.e., respectively for $S = T = 1$ and $S = T = -1$). This is true because F'''_{TS} has n - 2 factors of p, while F'_{TS} and F''_{TS} have n - 1 factors of p.

We cannot argue that (61) and (62) are always larger than (63), or even that they are necessarily larger than terms previously neglected (those with p_i^2 replaced by m^2). All we can say is that its largest possible value is larger. It is conceivable that in some problem (61) might vanish, but this would be expected to occur only in the simplest, most symmetrical problems (as it does for example, in $\pi^0 \to e^+ + e^-$), or under certain "accidental" conditions. In this connection it must be stressed that for a line with n interaction vertices there can be n! possible orderings (reduced, of course, when a virtual photon is emitted and reabsorbed by the same line) and if n is large the number of "accidents" required makes the vanishing of all the largest contributions unlikely in scattering problems.

In any case, when we evaluate the sum of F' or

F'' over all diagrams and find the expected order of magnitude, then we know we have a good approximation. We are, therefore, led to state

Rule II: The biggest contributions to the transition amplitude in the extreme relativistic limit are usually F'_{++} and F''_{--} (multiplied by their propagators). We call this approximate rule conservation of cohelicity.

Examining (61) and (62) we see that

$$|F'_{++}| = |F''_{--}|, \tag{64}$$

since the parts which do not depend on S and T are transformed into each other by the parity transformation ($p_i^+ \to p_i^-$, etc.).

Observe also that the factors a'_+, a_+ appearing on the right sides of Eqs. (61) and (62) are cancelled in the relativistic limit by the normalization to unit particle density, Eq. (16). That is, we have

$$\frac{a_+}{\sqrt{s_0}} = \sqrt{\frac{E + |\mathbf{p}|}{m}} \cdot \sqrt{\frac{m}{2E}} \approx 1. \tag{65}$$

In the next lecture we will consider techniques for evaluating matrix elements and cross-sections, using Rules I and II. I will conclude here by mentioning some earlier related work. Yennie, Ravenhall, and Wilson have used the massless approximation to the Dirac Equation to calculate the scattering of high energy electrons from nuclei. Sannikov[7] has considered a massless electrodynamics, with the fermion mass set zero even in the propagator functions, and has given several applications, as has also

Zel'dovich.[8] However, it may be objected that the mass enters in an essential way into radiative corrections, and Eriksson[2] has shown that the effective expansion parameter in potential scattering is

$$(\alpha/\pi)\ln(q^2/m^2)$$

where q^2 is the squared momentum transfer.

LECTURE III

1. - General Considerations

The Feynman rules give us the amplitude F_{fi} from which we obtain the cross-section:

$$\sigma = \frac{e^4}{\text{flux}} \prod_{\text{inc}\,i} (2E_i)^{-1} \int |F_{fi}|^2 \, 2\pi^4 \, \delta(p_f - p_i) \prod_{\text{fin}\,j} \left[\frac{d^3k}{2E(2\pi)^3} \right]_j \tag{66}$$

According to Rule II, the amplitude F is dominated in the high energy limit by F_{++} and F_{--} in a problem involving a single electron line and, in general, by the cohelicity conserving amplitudes. Since $|F_{++}| = |F_{--}|$, we calculate $|F_{++}|^2$ itself to get the spin average results for an unpolarized electron beam. Rule I tells us to ignore double corner diagrams, and Rule II tells us to calculate for the others (Eq. (58)), D^{-1} representing the propagators,

$$F'_{++} = m\, a'_+ a_+ (\chi_{t'}, (e_n^- \prod_{n-1}^{1} A_i)\chi_s) D^{-1} \tag{67}$$

where the χ_s' are "kinematic" spinors defined by Eqs. (45) and (46), and $A_i = p_{i+1}^+ e_i^-$ (Eq. (57)). Since the matrix

$$M^- = e_n^- \prod_{n-1}^{1} A_i \tag{68}$$

involves only σ-matrices it has the general form

$$M^- = M_0 - \underline{\sigma} \cdot \mathbf{M} \tag{69}$$

with

$$M_\mu = \frac{1}{2} \mathrm{Tr}\, [\bar{\sigma}_\mu M^-]. \tag{70}$$

Thus we need only to obtain the ++ matrix elements of 1 and σ.

Consider the situation described by Fig. 3, in

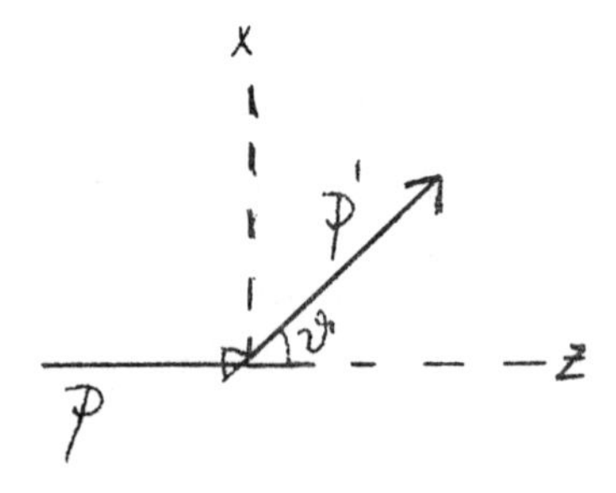

Fig. 3.

which an electron of momentum **p** in the Z-direction scatters to **p'** at angle ϑ in the x-z plane. The helicity spinor χ_+, corresponding to positive helicity on **p** is the positive eigenfunction of σ_z, namely $\binom{1}{0}$. On the other hand, the helicity operator $\boldsymbol{\sigma}\cdot\mathbf{p}/|\mathbf{p}|$ is transformed to σ_z by the unitary transformation

$$R = \exp\left(i\sigma_y \frac{\vartheta}{2}\right),$$

which therefore transforms the positive helicity spinor on $\mathbf{p}^+$, called χ'_+, into $\binom{1}{0}$. It follows that

$$\chi'_+ = e^{-i\sigma_y \frac{\vartheta}{2}} \binom{1}{0} = \begin{pmatrix} \cos\frac{\vartheta}{2} \\ \sin\frac{\vartheta}{2} \end{pmatrix}. \tag{71}$$

Thus

$$\left.\begin{aligned} {}_+\langle 1\rangle_+ &= {}_+\langle \sigma_z\rangle_+ = \cos\frac{\vartheta}{2} \\ {}_+\langle \sigma_x\rangle_+ &= -i\,{}_+\langle\sigma_y\rangle_+ = \sin\frac{\vartheta}{2} \end{aligned}\right\} \tag{72}$$

Consequently,

$${}_+\langle M^-\rangle_+ = (M_0 - M_z)\cos\frac{\vartheta}{2} - (M_x + iM_y)\sin\frac{\vartheta}{2}. \tag{73}$$

In cases where photon polarizations are to be summed or averaged over, a trace technique for the absolute square of the amplitude may be advantageous. The general rule ("backwards" rule, see Brown, reference 3) is

$$\sum_{s_1 s_2} | < \Omega_2 | M | \psi_1 > |^2 = \epsilon_1 \epsilon_2 \mathrm{Tr} [\overline{M} M], \qquad (74)$$

when the interaction is of the form given by Eq. (41). Of course, the double corners can still be eliminated by Rule I, since this is a general rule. The Matrix $\overline{M}$ is the matrix M with all the products of + and - operators written backwards with + and - interchanged, e.g., if

$$M = a^+ b^- \ldots y^+ z^- \qquad (75)$$

then

$$\overline{M} = z^+ y^- \ldots b^+ a^- . \qquad (76)$$

Projection operators to pick out positive energies are never necessary in this theory since the spinors describe only the spin, the particle-antiparticle character being given by the frequency of the plane wave part of the wave function. However, if we use the high energy approximation M^- in Eq. (68), which describes only the positive cohelicity conserving amplitude we must include appropriate cohelicity projection operator when summing over spins.
If N is the complete amplitude in the high energy limit, i.e. Eq. (68) with propagators included and summed over all diagrams, then

$$\sum_{s_1 s_2} |N_{12}|^2 = |N_{++}|^2 + |N_{--}|^2$$

$$= 2\,|N_{++}|^2 \tag{77}$$

$$= (2E_1E_2)^{-1}\,\mathrm{Tr}\,[p'^{+}\,N\,p^{+}\,\overline{N}\,].$$

In fact, referring to Eq. (67), and noting that

$$m\,a'_{+}\,a_{+} = (4E_1E_2)^{1/2} \tag{78}$$

we have just

$$|F_{++}|^2 = \mathrm{Tr}\,[p'^{+}\,N\,p^{+}\,\overline{N}\,]. \tag{79}$$

Because of the analogy between

$$\not{a}\,\not{b} + \not{b}\,\not{a} = 2\,\mathbf{a}\cdot\mathbf{b} \tag{80}$$

and

$$a^{+}b^{-} + b^{+}a^{-} = 2\,\mathbf{a}\cdot\mathbf{b} \tag{81}$$

the rules for traces of the type in Eq. (79) of Eq. (70) are the same as for the same traces with $\not{a} = a_\mu\,\gamma_\mu$, but there is one difference. The product of four different γ_μ is γ_5, whose trace is zero, but $\sigma_x\sigma_y\sigma_z = i$. Thus the trace of an odd number of σ matrices does not vanish, in general, and in using Eq. (70) a pseudoscalar term with imaginary coefficient may be present. For example,

$$\mathrm{Tr}[a^{+}b^{-}c^{+}d^{-}] = (\mathbf{a}\cdot\mathbf{b})(\mathbf{c}\cdot\mathbf{d}) + (\mathbf{a}\cdot\mathbf{d})(\mathbf{b}\cdot\mathbf{c})$$

$$- (\mathbf{a}\cdot\mathbf{c})(\mathbf{b}\cdot\mathbf{d}) + i\,\epsilon^{\mu\nu\sigma\tau} a_\mu b_\nu c_\sigma d_\tau . \qquad (82)$$

This usually causes no difficulty, when we recall that electrodynamics is parity conserving. These terms are important when the theory is applied to weak interactions.

These considerations will now be illustrated by examples.

2. - Example 1: Compton Scattering

a. - The Feynman diagrams and kinematic quantities are given in Fig. 4

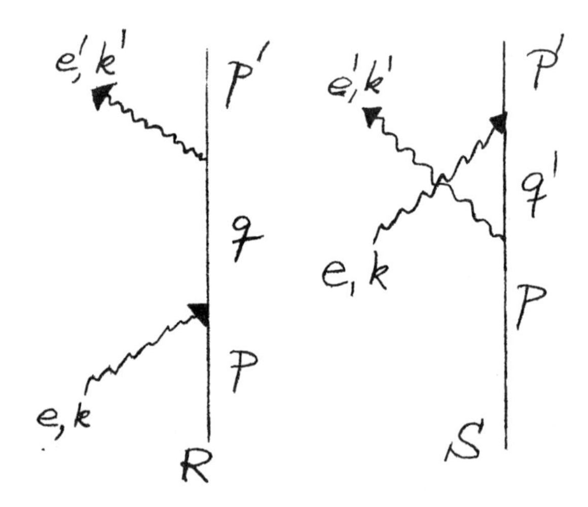

Fig. 4.

$$q = p + k = p' + k'$$

$$q' = p - k' = p' - k$$

$$m^2 \kappa \equiv m^2 - q^2 = -2p \cdot k = -2p' \cdot k'$$

$$m^2 \tau \equiv m^2 - q'^2 = 2p \cdot k' = 2p' \cdot k$$

The exact cross-section can be written in the invariant form:

$$d\sigma = \frac{2\pi \alpha^2}{m^2} \frac{\tau^2}{k^2} \left(\frac{d\tau}{\tau^2} + \frac{d\kappa}{\kappa^2} \right) u \tag{83}$$

with

$$u = 4\left(\frac{1}{\kappa} + \frac{1}{\tau}\right)^2 - 4\left(\frac{1}{\kappa} + \frac{1}{\tau}\right) - \left(\frac{\kappa}{\tau} + \frac{\tau}{k}\right), \tag{84}$$

to lowest order in α, and averaged over initial spins and polarizations.

Consider the C. M. system of the scattering (Fig. 5): In the system all energies and momenta are equal in the relativistic limit - call them ω. Thus,

$$m^2 \kappa = -4\,\omega^2$$

$$m^2 \tau = 2\omega^2 (1 + \cos \vartheta)$$

The result differential in angle becomes (put $d\kappa = 0$):

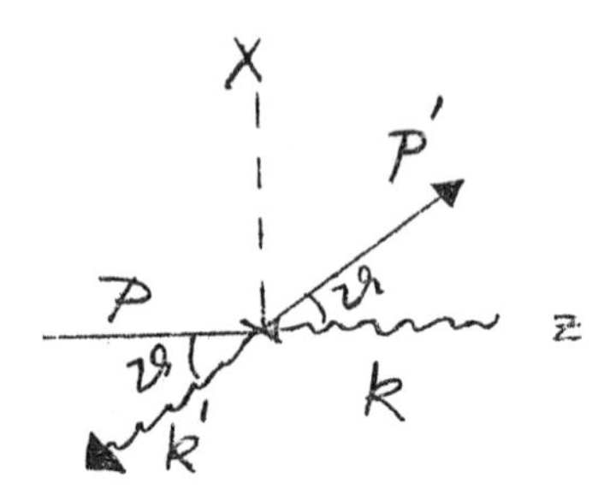

Fig. 5.

$$d\sigma = \frac{\alpha^2}{8\,\omega^2} \quad d\Omega\, u \tag{85}$$

with

$$\begin{aligned} u &\approx -\left(\frac{\kappa}{\tau} + \frac{\tau}{k} \right) = \frac{2}{1 + \cos\vartheta} + \frac{1 + \cos\vartheta}{2} \\ &= \cos^2\frac{\vartheta}{2} + \frac{1}{\cos^2\frac{\vartheta}{2}} \;. \end{aligned} \tag{86}$$

We have assumed, to obtain this result, $-\kappa, \tau \gg 1$ and therefore also $\omega(\pi - \vartheta) \gg 1$, so that a small region of backward angle scattering is excluded.

b. - Now let us compare the result calculated with the approximation method, Eq. (79). We have $(|F|^2 = 2u)$,

$$u = \frac{1}{4} \sum_{\text{pol}} \text{Tr}\,[p'^{+}(R + S)\, p^{+} (\overline{R} + \overline{S})] \tag{87}$$

with

$$R = -\frac{e^{1-}q^{+}e^{-}}{\kappa}, \quad S = -\frac{e^{-}q^{+}e^{1-}}{\tau} \tag{88}$$

so, suppressing, the alternating + and - superscripts,

$$u = \frac{1}{4\kappa^2}\,\mathrm{Tr}\,[p'\sigma_\mu q\sigma_\nu p\sigma_\nu q\sigma_\mu]$$

$$+\frac{1}{4\tau^2}\,\mathrm{Tr}\,[p'\sigma_\mu q'\sigma_\nu p\sigma_\nu q'\sigma_\mu]$$

$$+\frac{1}{2\kappa\tau}\,\mathrm{Tr}\,[p'\sigma_\mu q\sigma_\nu p\sigma_\mu q'\sigma_\nu]. \tag{89}$$

Using Eq. (81) and $\sigma_\mu a^+\sigma_\mu = -2a^+$, etc., the first trace is

$$4\mathrm{Tr}[qpqp'] = 4[2(p\cdot q)(p'\cdot q) - q^2(p\cdot p')] \cdot \mathrm{Tr}[1] \tag{90}$$

The second trace is the same with $q \to q'$, and the third trace is

$$-8\,(p\cdot p')(q\cdot q')\ \mathrm{Tr}\,[1], \tag{91}$$

where Tr[1] is, of course, equal to 2. Neglecting m^2/ω^2,

$$q^2 = 2p \cdot q = 2p' \cdot q = -\kappa$$

$$q'^2 = 2p \cdot q' = 2p' \cdot q' = -\tau$$

$$p \cdot p' = \frac{1}{2}(\kappa + \tau), \text{ and } q \cdot q' = 0.$$

Inserting these values we get

$$u = -\left(\frac{\tau}{\kappa} + \frac{\kappa}{\tau}\right).$$

in agreement with Eq. (86).

c. - For particular photon polarizations, it is more convenient to calculate the matrix elements directly. Choosing a gauge with spacelike polarization vector, Eq. (88) yields

$$M = R + S = -\frac{1}{k}\left[\hat{e}' q + \hat{e} + \frac{\kappa}{\tau}\hat{e} q'^{+}\hat{e}'\right]$$

with $\hat{e} = \boldsymbol{\sigma} \cdot \mathbf{e}$. In the C. M. system:

$$\frac{\kappa}{\tau} = -\frac{2}{1 + \cos\vartheta}, \quad q^{+} = 2\omega, \quad q'^{+}$$

$$= \omega[\sigma_x \sin\vartheta + \sigma_z(1 + \cos\vartheta)]$$

and

$$2\omega M = \hat{e}'\hat{e} - \hat{e}\sigma_x\hat{e}' \tan\frac{\vartheta}{2} - \hat{e}\sigma_z\hat{e}'. \tag{92}$$

Orthogonal transverse polarization vectors for the incoming photon are

$$\hat{e}^{(1)} = -\sigma_y, \quad \hat{e}^{(2)} = \sigma_x \tag{93}$$

and for the outgoing photon

$$e'^{(1)} = -\sigma_y, \quad e'^{(2)} = \sigma_x \cos\vartheta - \sigma_z \sin\vartheta. \tag{94}$$

Eq. (92) becomes, with notation $M^{ee'}$:

$$2\omega M'' = 1 + \sigma_x \tan\frac{\vartheta}{2} + \sigma_z$$

$$2\omega M^{22} = \cos\vartheta - i\sigma_y \sin\vartheta$$

$$- (\sigma_x \cos\vartheta - \sigma_z \sin\vartheta) \tan\frac{\vartheta}{2}$$

$$+ (\sigma_z \cos\vartheta + \sigma_x \sin\vartheta)$$

$$2\omega M^{12} = -i\sigma_z \cos\vartheta - i\sigma_x \sin\vartheta$$

$$+ (\sigma_y \cos\vartheta + i\sin\vartheta) \tan\frac{\vartheta}{2}$$

$$+ i\cos\vartheta - \sigma_y \sin\vartheta \tag{95}$$

$$2\omega M^{21} = i\sigma_z + \sigma_y \tan\frac{\vartheta}{2} - i.$$

Now use Eq. (72) to obtain the ++ matrix element

$$\left.\begin{aligned} 2\omega \langle M^{11} \rangle &= \frac{1}{\cos\frac{\vartheta}{2}} + \cos\frac{\vartheta}{2} = 2\omega \langle M^{22} \rangle \\ 2\omega \langle M^{12} \rangle &= -i \sin\frac{\vartheta}{2} \tan\frac{\vartheta}{2} = -2\omega \langle M^{21} \rangle \end{aligned}\right\} . \tag{96}$$

Checking again the polarization average, we get

$$\frac{a_+^2 a_+'^2}{2} \left(2 |\langle M^{11} \rangle|^2 + 2 |\langle M^{12} \rangle|^2 \right)$$

$$= \left(\frac{1}{\cos\frac{\vartheta}{2}} + \cos\frac{\vartheta}{2} \right)^2 + \sin^2\frac{\vartheta}{2} \tan^2\frac{\vartheta}{2}$$

$$= \frac{1}{\cos^2\frac{\vartheta}{2}} + \cos^2\frac{\vartheta}{2} + \frac{2\cos^2\frac{\vartheta}{2} + (1 - \cos\frac{\vartheta}{2})^2}{\cos^2\frac{\vartheta}{2}} \tag{97}$$

$$= 2\left(\frac{1}{\cos^2\frac{\vartheta}{2}} + \cos^2\frac{\vartheta}{2} \right) = 2u = |F|^2 .$$

3. - Example 2. Relativistic Rutherford Scattering

For the first order result we merely note that (Fig. 6):

$$M^- = M_0 = \frac{4\pi}{Q^2} Z e^2 \tag{98}$$

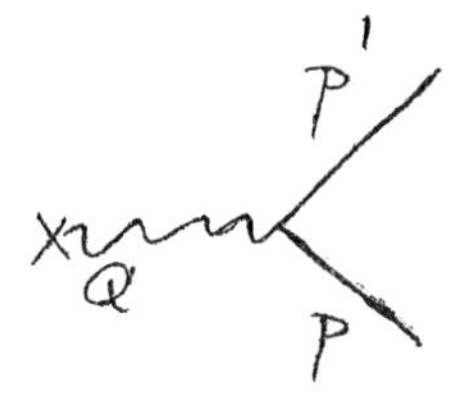

Fig. 6.

with $Q^2 = 4p^2 \sin\frac{\vartheta}{2}$. This result is the well-known Fourier transform of the Coulomb potential.

Thus;

$$M_{++}^{(1)} = M_0 \cos\frac{\vartheta}{2} \tag{99}$$

by Eq. (73).

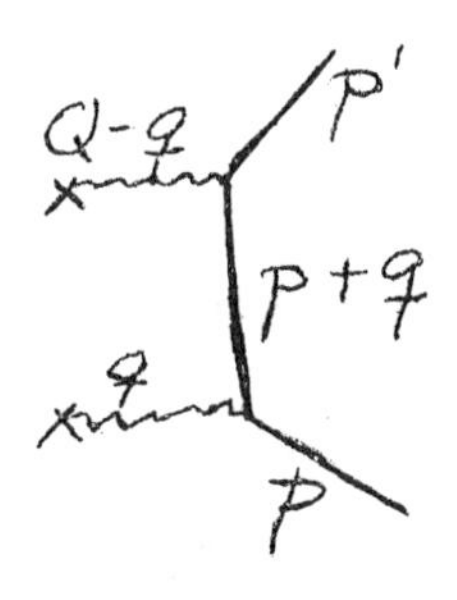

Fig. 7.

For the second order result, Fig. 7 gives

$$M_{++}^{(2)} = \frac{(Q^2 M_0^2)^2}{(2\pi)^3} \int \frac{{}_{+}\langle p^+ + q^+ \rangle_{+} \, d^3q}{(q^2 + \lambda^2)[(Q - q)^2 + \lambda^2](q^2 + 2p\cdot q)}$$

$$= \frac{(Q^2 M_0^2)^2}{(2\pi)^3} \left[2E \cos\frac{\vartheta}{2} I_0 + {}_{+}\langle \sigma_j \rangle_{+} I_j \right], \tag{100}$$

as

$${}_{+}\langle p^+ \rangle_{+} \approx 2E \, {}_{+}\langle 1 \rangle_{+} = 2E \cos\frac{\vartheta}{2} \, .$$

The integrals required are

$$(I_0, I_j) \equiv \lim_{\lambda \to 0} \int \frac{(1, q_j) \, d^3q}{(q^2 + \lambda^2)[(Q - q)^2 + \lambda^2](q^2 + 2p\cdot q)} \, . \tag{101}$$

The quantity M_0, Eq. (98) gives the Rutherford cross-section. Comparing Eq. (99), we will therefore write the correction as

$$R = \cos^2\frac{\vartheta}{2} \left(1 + \frac{2\mathrm{Re}\, M_{++}^{(1)} M_{++}^{(2)}}{|M_{++}^{(1)}|^2}\right). \tag{102}$$

We note that the integral I_0 contains a $\ln\lambda$, where λ is a screening parameter. This combines with a similar term in I_j to give the first order term in the expansion of the infinite Coulomb phase factor,[9] so that we ignore these terms. The remaining relevant term in I_j is

$$I'_j = \frac{1}{2} (p + p')_j \frac{\pi^3}{pQ^2} \frac{(1 - \sin\frac{\vartheta}{2})}{\cos\frac{\vartheta}{2} \operatorname{ctn}\frac{\vartheta}{2}} . \tag{103}$$

The factor

$$\frac{1}{2} {}_{+}\langle (\mathbf{p} + \mathbf{p}') \cdot \boldsymbol{\sigma} \rangle_{+} = E \cos\frac{\vartheta}{2} ,$$

and then Eq. (102) becomes

$$R_{ER} = \cos^2\frac{\vartheta}{2} + \pi Z e^2 \sin\frac{\vartheta}{2} (1 - \sin\frac{\vartheta}{2}) \tag{104}$$

to be compared with the correction factor given by McKinley and Feshbach[10]

$$R = 1 - \beta^2 \sin^2\frac{\vartheta}{2} + \pi Z e^2 \beta \sin\frac{\vartheta}{2} (1 - \sin\frac{\vartheta}{2}). \tag{105}$$

FOOTNOTES

1) For the experimental situation see the 1966 Berkeley International High Energy Physics Conference report.
2) N. Meister and D. R. Yennie, Phys. Rev. 130,

1210 (1963). See also: D. R. Yennie, S. C. Frautschi and H. Suura, Ann. of Phys. 13, 379 (1961) and K. E. Eriksson, Nuovo Cimento 19, 1044 (1961).

3) L. M. Brown, Phys. Rev. 111, 957 (1958)
M. Tonin, Nuovo Cim. 14, 1108 (1959)
H. Pietschmann, Acta Phys. Austriaca 14, 63 (1961)
L. M. Brown, Lectures in Theoretical Physics vol. IV (International Publishers, New York 1962).

4) L. M. Brown and Ik-Ju Kang, Nuovo Cimento, 52A, 210 (1967) and Ik-Ju Kang, Ph. D. Dissertation, Northwestern University, USA (1962).

5) E. g., S. Gasiorowicz, Elementary Particle Physics (John Wiley and Sons, Inc., New York 1966).

6) D. R. Yennie, Ravenhall and Wilson, Phys. Rev. 95, 500 (1954).

7) S. S. Sannikov, JETP 40, 237 (1961); Soviet Physics JETP 13, 163 (1961).

8) Ya. B. Zeldovich, JETP 41, 912 (1961); Soviet Physics JETP 14, 651 (1962).

9) R. H. Dalitz, Proc. Roy. Soc. A206, 509 (1951).

10) W. A. McKinley Jr., and H. Feshbach, Phys. Rev. 74, 1759 (1948).

THE CURRENT ALGEBRA APPROACH TO ELEMENTARY PARTICLE PHYSICS

Curtis G. Callan, Jr.

THE CURRENT ALGEBRA APPROACH TO ELEMENTARY PARTICLE PHYSICS

Curtis G. Callan, Jr.
Harvard University

I. INTRODUCTION

By far the most interesting developments in elementary particle physics in recent years are those involving the so-called algebra of currents. They include the Adler-Weisberger formula for g_A, calculations of pion-nucleon scattering lengths, and various predictions for nonleptonic and semileptonic K-meson decays. If one considers the ensemble of results, and their remarkable agreement with experiment, it is evident that a very powerful insight into the physics of strongly interacting particles has been obtained. In these lectures I will try to give a concise introduction to the basic ideas of the subject and explain a few of the more interesting applications. Since the current algebra is not really a well-defined theory, but a set of extremely useful recipes which must be used with care, the best that I can hope to do in the short time available to me is to give a general idea of how things are done, leaving it to the reader to study the applications in more detail.

Although the most interesting features of this scheme concern the strong interactions, the basic equations concern the currents which are measured in the semi-leptonic weak interactions. Therefore, we shall begin with a review of the phenomenology of the weak interactions. By now considerable evidence has accumulated that the weak interactions can be described by the simple Lagrangian

$$L_w = \frac{G}{\sqrt{2}} J^{\mu} J^{+}_{\mu}$$

$$J_{\mu} = \overline{\psi}_e \gamma_{\mu} (1 + \gamma_5) \psi_{\nu_e} + \overline{\psi}_{\mu} \gamma_{\mu} (1 + \gamma_5) \psi_{\nu_{\mu}}$$

$$+ \cos\theta \, (V_{\mu}^{\pi^+} + A_{\mu}^{\pi^+}) + \sin\theta \, (V_{\mu}^{K^+} + A_{\mu}^{K^+}).$$

The objects V_{μ} and A_{μ} are, respectively, vector and axial-vector operators constructed out of the fields of strongly-interacting particles. They are conventionally called the vector and axial-vector currents of weak interactions. The indices π^+ and K^+ indicate that the currents have the internal quantum numbers of the corresponding particles: $(I, I_3, S)_{\pi^+} = (1, 1, 0)$, $(I, I_3, S)_{K^+} = (\frac{1}{2}, \frac{1}{2}, 1)$. The Cabibbo angle, θ, which is found to be about .2 radians, expresses the fact that decays in which the strangeness of the hadrons changes by one unit seem to be intrinsically weaker than those in which the strangeness does not change.

The coupling constant G is so weak that it makes sense for most purposes to use the above Lagrangian only to first order. Therefore if we wish to study the decay of hadron A into hadron B with the emission of electron and neutrino

$$A \to B + e^- + \bar{\nu}_e$$

we easily find that the decay matrix element is

$$M = \frac{G}{\sqrt{2}} \bar{u}_e \gamma^\mu (1 + \gamma_5) v_{\bar{\nu}} \langle B | (V_\mu^{\pi^+} + A_\mu^{\pi^+}) | A \rangle \cos\theta$$

if $S_A = S_B$, and

$$M = \frac{G}{\sqrt{2}} \bar{u}_e \gamma^\mu (1 + \gamma_5) v_{\bar{\nu}} \langle B | (V_\mu^{K^+} + A_\mu^{K^+}) | A \rangle \sin\theta$$

if $S_B = S_A + 1$. Therefore, any semi-leptonic decay of this type is known if we know the matrix elements between hadronic states of the currents V_μ and A_μ. This of course is where the difficulty lies, since the strong interactions enter into such matrix elements in all their complexity.

Nonetheless, a certain number of principles have been discovered which allow us to make useful predictions. In the first instance, the current $V_\mu^{\pi^+}$ seems to be identical to the corresponding charge component of the isospin current. This means that matrix elements of $V_\mu^{\pi^+}$ between different members of an isospin multiplet are completely known in terms of Clebsch-Gordon coefficients of the isospin group. Therefore, decays such as $n \to p + e^- + \bar{\nu}$ have known rates. Furthermore, since the electromagnetic current is the sum $V_\mu^{\pi^0} + \frac{1}{2}B_\mu$, B_μ being the isotopic spin zero baryon number current, we may hope by experiments involving photons to determine matrix elements of $V_\mu^{\pi^0}$. But then isotopic spin invariance allows us to translate that into information about $V_\mu^{\pi^+}$. This gives us a general connection between electromagnetic and semi-leptonic weak processes.

Further information of this type is obtainable by considering the extension of isospin symmetry, SU_3. The simplest assumption one might make is that the observable vector currents

$$(V_\mu^{\pi^+},\ V_\mu^{K^+},\ (V_\mu^{\pi^+})^+ = V_\mu^{\pi^-},\ (V_\mu^{K^+})^+ = V_\mu^{K^-})$$

are all members of the same SU_3 multiplet. The smallest available multiplet containing operators of these quantum numbers is the octet, so we shall assume that the vector currents are all members of the same octet. We may make the same assumption for the axial vector currents. Since $V_\mu^{\pi^+}$ is identical with the charge raising component of the isospin current, which is one of the generators of SU_3, the other vector currents must also be generators of SU_3. Therefore we arrive at the simple scheme of identifying the vector currents of the weak interactions with SU_3 generators and asserting that the axial-vector currents are likewise members of an octet. This has very strong consequences in the limit that SU_3 symmetry is exact: from some small number of parameters we can determine all the matrix elements of the currents between the various states of an SU_3 multiplet. These consequences appear to be well-verified for the baryon octet, indicating that the scheme makes a certain amount of sense.

On the other hand, we know that SU_3 is not an exact symmetry, and that sometimes the violations can be considerable. The question then arises of what features of the scheme we have just elaborated can be retained in the real world. Let us imagine that the Hamiltonian is

$$H_0 + \lambda H_1$$

where H_0 is invariant under SU_3 and H_1 breaks the symmetry. The invariance of H_0 means that there exists a group of unitary transformations

$$\psi'_\alpha = U_{\alpha\beta} \psi_\beta$$

on the set, ψ_α, of fields out of which H_0 is constructed leaving H_0 invariant. If the invariance group is SU_3, there are a set of eight vector currents V^a_μ, a = 1, ..., 8, which may be constructed out of the fields, having the following properties:

a. $F^a(t) = \int d\overline{x}\, V_0{}^a(\overline{x}, t)$ are the generators of the symmetry transformations:

if $U_t(\vec{\lambda}) = \exp(i \lambda^a F^a(t))$, then

$$U_t(\overline{\lambda})\, \psi_\alpha(t)\, U_t{}^{-1}(\overline{\lambda}) = U_{\alpha\beta}\, \psi_\beta(t)$$

where the $U_{\alpha\beta}$ are some matrix representation of the SU_3 group.

b. The currents satisfy simple equal-time commutation relations

$$[V_0^a(\bar{x}, t), V_0^b(0, t)] = i\delta(\bar{x}) f_{abc} V_0^c(t)$$

when f_{abc} are the structure constants of the SU_3 algebra.

c. If $0^a(x)$ is an SU_3 octet of local operators, it satisfies the simple set of equal-time commutation relations

$$[V_0^a(\bar{x}, t), 0^b(0, t)] = i\delta(\bar{x}) f_{abc} 0^c(t).$$

d. If the parameter $\lambda = 0$, the currents V_μ^a are conserved and the generators F^a are independent of time.

It is the latter property that we need in order to conclude that SU_3 multiplets of particles having the same mass etc., exist. However, in the real world, the parameter λ must be in fact non-zero since none of the predictions of pure SU_3 are exactly true. On the other hand, nothing prevents us from assuming that the currents V_μ^a defined by the SU_3-invariant part of the hamiltonian an identical with the vector currents of the weak interactions. This would be the natural extension of the idea that $V_\mu^{\pi^+}$, the non-strangeness-changing weak current, is identical with one of the isospin currents. In that case we may retain the equal-time commutation relations b). By the same token, what remains of the idea that the axial-vector weak currents are members of an octet are the equal-time commutation relations c):

$$[V_0^a(\bar{x}, t), A_0^b(0, t)] = i\delta(\bar{x}) f_{abc} A_0^c(0, t).$$

It is to be emphasized that only equal-time commutators are known in this limit - we no longer have exact SU_3 symmetry to tell us that the generators F^a are independent of time.

These equal-time commutators are of course not sufficient to determine the matrix elements of the currents. As we shall see later on, they enable us to establish sum rules which permit the calculation of certain quantities in terms of others.

We can say, however, that the relation

$$[V_0^a(\bar{x}, 0), V_0^b(0)] = i\delta(\bar{x}) f_{abc} V_0^c(0),$$

since it is non-linear, suffices to determine the scale of the matrix elements of the vector current. In contrast, the relation

$$[V_0^a(\bar{x}, 0), A_0^b(0)] = i\delta(\bar{x}) f_{abc} A_0^c(0),$$

being linear in the axial current does not determine the scale of matrix elements of A_μ^a. In order to have a handle on this we should have to consider the commutator

$$[A_0^a(\bar{x}, 0), A_0^b(0)],$$

about which we have no a priori information.

We may, however, make reasonable conjectures about this sort of commutator by considerations similar to those which led us to the equal-time commutation relations of the vector current with itself.

Let us make the bold supposition that the SU_3 invariant piece of the Hamiltonian, H_0, may be written

$$H_0 = H_{00} + \eta H_{10}$$

where H_{00} is invariant under an even broader symmetry group than SU_3. Then we will find n currents,

$$J_\mu^\alpha, \ (\alpha = 1, \ldots, n),$$

where n is the number of generators of the group, such that

$$F^\alpha = \int d\bar{x} \ J_0^\alpha$$

act as the generators of the group transformations on the fundamental fields, and such that J_0^α satisfy equal-time commutation relations determined by the structure constants of the group. Clearly we want J_μ^α to contain V_μ^a, A_μ^a since we are interested in the mutual commutation relations of precisely these objects. The simplest symmetry group containing 8 vector and 8 axial-vector generators is $SU_3 \times SU_3$ - which has, as a matter of fact, <u>only</u> those generators The equal-time commutation relations among the currents required by the structure of the group are

$$[V_0^a(0, \bar{x}), V_0^b(0)] = i\delta(\bar{x}) f_{abc} V_0^c(0)$$

$$[V_0^a, A_0^b] = i\delta(\overline{x}) f_{abc} A_0^c$$

$$[A_0^a, A_0^b] = i\delta(\overline{x}) f_{abc} A_0^c .$$

The first two commutators are familiar to us, but the last one is entirely new. It is non-linear in the axial-vector currents and therefore should serve to define their scale, which is precisely what we should like to do. It is interesting that within the framework of a quark model it is possible to write down Lagrangians having the structure we postulated above. H_{00}, in general, contains the kinetic part of the Lagrangian plus some interaction term, while H_{10}, the piece which breaks the $SU_3 \times SU_3$ symmetry, is just the quark mass term.

Therefore, we may reasonably take the same position with respect to $SU_3 \times SU_3$ symmetry as we previously took with respect to SU_3 itself. We know that the symmetry is extremely badly broken, but we may retain some vestiges of it by identifying the vector and axial-vector currents of the weak interactions with the generators of $SU_3 \times SU_3$, which, although they are not conserved, have the simple equal-time commutation relations which we have just written down. The great advantage of this speculative scheme is that it is non-linear in both the vector and axial-vector currents, and hence should allow us to determine the scale of both.

There are other properties of the currents which we must establish before we can extract the most far-reaching applications of this scheme of equal-time commutation relations. However, it is probably worthwhile to pause here and show that with the properties we have already proposed we can obtain interesting sum rules for measureable quantities.

We shall use a trick as old as non-relativistic quantum mechanics to obtain a sum rule for high-energy neutrino reactions. If we have a commutator $[A, A^{+}] = B$, then by taking matrix elements and inserting a complete set of intermediate states, we get the sum rule

$$\sum_{s} \{ |\langle s|A^{+}|a\rangle|^{2} - |\langle s|A|a\rangle|^{2} \} = \langle a|B|\rangle .$$

Its usefulness depends, of course, on how easy it is to measure the matrix elements involved in the sum.

According to the theory of semi-leptonic interactions the matrix elements for reactions of the type $\nu + p \to s + l^{-}$, when s is any state of strongly interacting particles, is

$$A(\nu + p \to l^{-} + s) = \frac{G}{\sqrt{2}} u_{l} \gamma^{\mu} (1 + \gamma_{5}) u_{\nu} \langle s|(V_{\mu}^{\pi^{-}} + A_{\mu}^{\pi^{-}})|p\rangle$$

$$A(\bar{\nu} + p \to l^{+} + s) = \frac{G}{\sqrt{2}} \bar{\nu}_{\bar{\nu}} \gamma^{\mu} (1 + \gamma_{5}) v_{\bar{l}} \langle s|(V_{\mu}^{\pi^{+}} + A_{\mu}^{\pi^{+}})|p\rangle$$

Therefore, in these reactions we can hope to measure

$$\langle s | (V_\mu^{\pi^\pm} + A_\mu^{\pi^\pm}) | p \rangle$$

where s is any state of strongly interacting particles allowed by the conservation laws. In particular, if we neglect the lepton mass, this means that the momentum transfer $(p - s)^2$ must be negative.

To obtain a sum rule, we shall start with the commutator

$$[J_0^+ (0, \bar{x}), J_0^- (0)] = 2\,\delta(\bar{x})\, J_0^0(0)$$

$$J_\mu^\pm = V_\mu^{\pi^\pm} + A_\mu^{\pi^\pm}$$

$$J_\mu^0 = V_\mu^{\pi^0} + A_\mu^{\pi^0}$$

and sandwich it between identical proton states of momentum $\bar{p}$:

$$\sum_s \Big\{ e^{i(\bar{P}_s - \bar{p}) \cdot \bar{x}} |\langle s | J_0^- | p \rangle|^2 - e^{-i(\bar{P}_s - \bar{p}) \cdot \bar{x}} |\langle s | J_0^+ | p \rangle|^2 \Big\}$$

$$= 2\delta(\bar{x})\, \langle p | J_0^0 (0) | p \rangle .$$

We observe that since $V_0{}^{\pi^0}$ is just the third component of isospin, we may write $\langle p | V_0{}^{\pi^0} | p \rangle = \frac{1}{2}$. Similarly $\langle p | A_0{}^{\pi^0} | p \rangle = \frac{1}{2} g_A h\, v_p$, where h is the proton's helicity, v_p its velocity, and g_A the axial-vector renormalization constant. We shall find it convenient to average our sum rule over proton spin (thereby getting rid of the term proportional to g_A) and to take a Fourier transform, giving

$$\frac{1}{2} \sum_{\substack{\text{proton}\\ \text{spin}}} \sum_{s} \Big\{ (2\pi)^3 \delta(\overline{P}_s - \overline{p} - \overline{k}) \, | \langle s | \overline{J}_0 | p \rangle |^2$$

$$- (2\pi)^3 \delta(\overline{p}_s - \overline{p} + \overline{k}) \, | \langle s | J_0^+ | p \rangle |^2 \Big\}$$

$$= 1.$$

In order to transform this into a more convenient form we insert into the sum the identity

$$1 = \int \frac{dM^2}{2\pi} \cdot 2\pi \, \delta(P_s^2 - m^2)$$

and observe that

$$\delta(P_s^2 - M^2) \, \delta(\overline{P}_s - \overline{p} - \overline{k}) = \frac{1}{2E} \delta(P_{s_0} - E) \, \delta(\overline{P}_s - \overline{p} - \overline{k})$$

$$= \frac{1}{2E} \, \delta^{(4)}(P_s - p - k)$$

where

$$E = M^2 + \overline{P}_s^2 , \; k_0 = E - p_0 .$$

The sum rule becomes

$$1 = \int \frac{dM^2}{2\pi} \frac{1}{2E} \sum_s \left\{ (2\pi)^4 \delta(P_s - p - k) \, |\langle s | J_0^- | p\rangle|^2 \right.$$

$$\left. - (2\pi)^4 \delta(P_s - p - k) \, |\langle s | J_0^+ | p\rangle \right\}$$

The four-vectors k, k are defined by

$$k = (k_0, \bar{k}) \quad (p + k)^2 = M^2$$

$$k = (k_0, -\bar{k}) \quad (p + k)^2 = M^2,$$

and the average over proton spin is henceforth implicit.

The sum

$$\sum_s (2\pi)^4 \delta(P_s - p - k) \, |\langle s | J_0^{\pm} | p\rangle|^2$$

is obviously closely related to the cross-section for $\nu + p \to l +$ (any strongly interacting state) with lepton momentum transfer fixed at k. In fact, the squared S-matrix element for such a process is

$$|M|^2 = \frac{G^2}{2} \bar{u}_1 \gamma_\mu (1 + \gamma_5) v_\nu \bar{v}_\nu \gamma_\nu (1 + \gamma_5) u_1$$

$$\times \sum_s (2\pi)^4 \delta(P_s - p - k) \langle p | J_\nu^- | s\rangle \langle s | J_\mu^+ | p\rangle$$

For convenience, make the definition

$$\frac{T^{\pm}_{\mu\nu}}{2 p_0} = \sum_s (2\pi)^4 \delta(p_s - p - k) \langle p | J^-_\nu | s \rangle \langle s | J^+_\mu | p \rangle .$$

Then to test the sum rule we need to know T_{00}.

Since $T_{\mu\nu}$ is a Lorentz tensor formed out of p and k, its most general form is

$$T_{\mu\nu} = \alpha\, \delta_{\mu\nu} + \beta\, p_\mu p_\nu + \gamma\, \epsilon_{\mu\nu\lambda\sigma} p_\lambda k_\sigma + \delta k_\mu k_\nu$$

$$+ \epsilon p_\mu k_\nu + K p_\nu k_\mu ,$$

when $\alpha = \alpha\,(M^2, k^2)$, etc. On the other hand

$$k_\mu \bar{u}_1 \gamma_\mu (1 + \gamma_5) u_\nu = m_1 \bar{u}_1 (1 + \gamma_5) v_\nu$$

since $k = p_1 - p_\nu$. Since the lepton mass is negligible at the energies at which one does this kind of experiment, the contribution of δ, ϵ, K to $|M|^2$ is negligible. This means that one can only hope to measure α, β and γ and that we cannot measure T_{00} completely!

Fortunately, there is a way around this difficulty. Consider $T_{00}(k, p)$ for fixed $M^2 = (k + p)^2$. Then, keeping $\bar{k}$ fixed, let $\bar{p}$ become large:

$$k_0 = \sqrt{M^2 + (\bar{p} + \bar{k})^2} - \sqrt{m^2 + p^2} \rightarrow |\bar{k}| \cos\theta$$

$$\left(\cos\theta = \frac{\bar{p} \cdot \bar{k}}{|\bar{p}|\,|\bar{k}|}\right)$$

$$k^2 \to -|\bar{k}|^2 \sin^2\theta$$

$$E \to p_0 \to |\bar{p}|$$

$$\frac{T_{00}(k, p)}{E_{p_0}} \to \beta(M^2, -k^{-2} \sin^2\theta).$$

But T_{00}/E_{p0} is precisely what appears in the sum rule! Therefore, if we assume that it is permissible to interchange

$$\int dM^2 \text{ and } \lim_{\bar{p}\to\infty} ,$$

we obtain the simple result

$$1 = \int \frac{dM^2}{8\pi} \left(\beta^-(M^2, k^2) - \beta^+(M^2, k^2)\right)$$

with $k^2 < 0$ and fixed in the integration. But $\beta^\pm$ is a quantity which we <u>can</u> measure in high energy neutrino scattering experiments on unpolarized targets. Therefore, in principle we can test this sum rule, although it will surely be some time before the necessary data are available.

In order to obtain results which are more readily comparable with experiment, we shall have to add some new information to the equal-time commutation relations which have been proposed. Let us consider the question of the divergences of the currents we have been studying. The current $v_\mu^{\pi+}$, since it is identical with the isospin current, must be conserved, at least in the absence of electromagnetism. On the other hand v_μ^{k+} may not be conserved since that would mean that SU3 symmetry is good. What about the axial currents? Let us first consider $A_\mu^{\pi+}$ and ask whether the equation $\partial_\mu A_\mu^{\pi+} = 0$ is possible. We observe that the decay $\pi^- \rightarrow e^- \bar{\nu}$ is determind by the matrix element

$$\langle 0 | A_\mu^{\pi^+} | \pi^- (q) \rangle$$

which has the general form $f_\pi q_\mu$. But then

$$- \langle 0 | \partial_\mu A_\mu^{\pi^+} | \pi^- \rangle = - i f_\pi q^2 = - i \mu_\pi^2 f_\pi,$$

or $f_\pi = 0$, if $A_\mu^{\pi+}$ is conserved. But this is impossible because $f_\pi = 0$ would mean that the pion cannot decay.

Let us change our point of view slightly and consider a model theory in which the physical mass of the pion is zero. It is still true that

$$\langle 0 | A_\mu^{\pi^+} | \bar{\pi} (q) \rangle = f_\pi q_\mu,$$

but since $q^2 = \mu^2{}_\pi = 0$, it is automatically true that

$$\langle 0 | \partial_\mu A_\mu^{\pi^+} | \pi^- \rangle = 0$$

independent of the magnitude of f_π. Therefore, in such a theory, the axial current may be conserved and the pion still have non-vanishing weak interactions. If one studies the question in more detail one finds that as long as the physical mass of the pion is zero, the equation $\partial_\mu A_\mu^{\pi^+} = 0$ does not imply any qualitative change in pion interactions from a more general theory.

We may ask what relevance this has to the real world, where the pion mass is <u>not</u> zero. The crucial point is that on the scale of typical strong-interaction energies and masses, the pion mass is quite small, and we may reasonably hope that the world in which $\mu_\pi = 0$ and $\partial_\mu A_\mu^\pi = 0$ is a good approximation to the real world. We shall wait until we come to specific applications to discuss how this comparison might be made as well as its limits of validity.

For the moment let us study the model world in which $\mu_\pi = 0$ and $\partial_\mu A_\mu^\pi = 0$, and see what conclusions may be drawn. First of all, consider the matrix element

$$M_\mu = \langle \beta | A_\mu^{\pi^+} | \alpha \rangle, \quad q_\mu = (P_\beta - p_\alpha)_\mu$$

where α, β are any state of strongly interacting particles. Since the axial current is conserved, we have the simple equation

$$iq^\mu M_\mu = 0.$$

On the other hand, let us consider the limit $q \to 0$. In the diagrammatic expansion of the matrix element M_μ, only those terms which have a pole as $q \to 0$ will survive in

$$\lim_{q \to 0} q^\mu M_\mu \, .$$

Such pole terms are easily classified as follows:

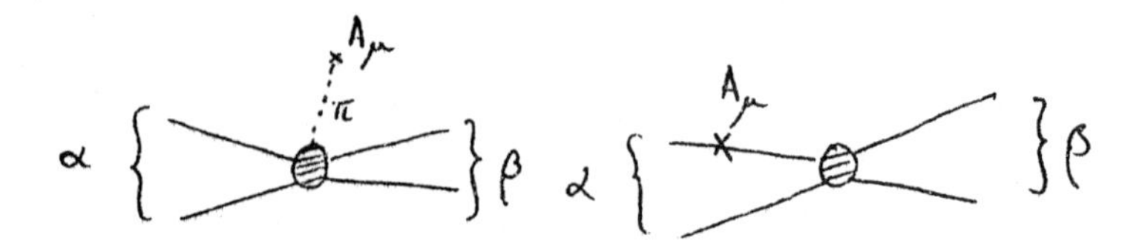

In the first case, the axial current creates a pion out of vacuum, in the second case it acts on an external baryon line, and in either case, a propagator which goes on the mass shell in the limit $q \to 0$ is present. The first sort of diagram contributes a term

$$f_\pi \langle \beta\pi(0) | \alpha \rangle$$

to

$$\lim_{q \to 0} iq^\mu M_\mu$$

while the second sort of diagram contributes a term completely determined by the matrix element $\langle \beta | \alpha \rangle$ (but which contains some relatively complicated kinematical factors.) Therefore, the fact that the axial vector current is conserved allows us to derive a host of '' low-energy theorems'' which connect an arbitrary process $\alpha \to \beta$ to $\alpha \to \beta + \pi$ (zero energy). This sort of relation will turn out to be extremely useful.

We can also make a slight generalization of this idea which allows us to make use of the equal-time commutation relations which were proposed earlier. Consider the quantity

$$M_{\mu\nu} = \int dx\, e^{ik \cdot x} \langle\beta|T(A_\mu^\pi(x)\, J_\nu(0))\,|\alpha\rangle,$$

where J_ν is an arbitrary current operator. Then we have the identity

$$-ik^\mu M_{\mu\nu} = \int dx\, (-\frac{\partial}{\partial x_\mu}\, e^{ik \cdot x})$$

$$\langle\beta|T(A_\mu^\pi(x)\, J_\nu(0)\,|\alpha\rangle$$

$$= \int dx\, e^{ik \cdot x} \frac{\partial}{\partial x_\mu}$$

$$\langle\beta\,|T(A_\mu^\pi(x)\, J_\nu(0))\,|\alpha\rangle$$

$$= \int dx\, e^{ik \cdot x}\, \delta(x_0)$$

$$\langle\beta|\,[A_0^\pi(x),\, J_\nu(0)]\,|\alpha\rangle.$$

To obtain the last line, we use the identity

$$\frac{\partial}{\partial x_\mu} T\,(A_\mu(x) J_\nu(0)) = T(\partial^\mu A_\mu(x) J_\nu\,(0))$$

$$+\ \delta\,(x_0)\,[A_0(x),\, J_\nu(0)]$$

along with the assumed conservation of A_μ^π. We now see that if J_μ is some combination of the weak currents, whose equal-time commutation relations we know, we will be able to write

$$\delta(x_0)\,[A_0^{\pi}(x),\; J_\nu(0)] = \delta^{(4)}(x)\,\bar{J}_\nu(0)$$

where $\bar{J}$ is some new combination of the weak currents. Therefore

$$-ik^\mu M_{\mu\nu} = \langle \beta | \bar{J}_\nu | \alpha \rangle .$$

If we now consider the limit $k \to 0$, only those contributions to $M_{\mu\nu}$ which have poles in k will survive. As before, these can be divided into two classes: 1) the axial current creates a pi-meson out of vacuum, giving a contribution $f_\pi \langle \beta\pi(0) | J_\nu | \alpha \rangle$ to $-ik^\mu M_{\mu\nu}$, 2) the axial current tacks on to an external baryon line, giving a contribution entirely determined by $\langle \beta | J_\nu | \alpha \rangle$. Things are considerably simplified if β and α are purely mesonic states, in which case only the first class of diagrams exists. Then we can conclude that

$$f_\pi \langle \beta\pi(0) | J_\nu | \alpha \rangle = \langle \beta | \bar{J}_\nu | \alpha \rangle ,$$

giving us an interesting connection between two different weak decays.

We are now in a position to see how this scheme can be used in a physically interesting situation. We consider the complex of semileptonic decays of K-mesons:

1) $K \to l\nu$

2) $K \to \pi l \nu$

3) $K \to 2\pi l \nu$,

concerning which there is a considerable amount of experimental information. To be specific, we shall first consider $K^+ \rightarrow \pi^0 + e^+ + \nu$. The general theory of semi-leptonic interactions says that the matrix element for this decay has the form

$$M = \frac{G}{\sqrt{2}} \bar{u}_\nu \gamma^\mu (1 + \gamma_5) v_e$$

$$\langle \pi^0(p') | (V_\mu^{K^-} + A_\mu^{K^-}) | K^+(p) \rangle .$$

From Lorentz invariance and parity conservation in strong interactions

$$\langle \pi^0 | A_\mu^{K^-} | K^+ \rangle = 0$$

$$\langle \pi^0 | V_\mu^{K^-} | K^+ \rangle = f_+ (p + p)_\mu + f_- (p - p')_\mu$$

where $f_\pm$ are functions of a single variable which we may take to be $p \cdot p'$. The physical region of this variable is

$$M_K m_\pi \leq p \cdot p' \leq \frac{M_K^2 + m_\pi^2}{2}$$

or, if $m_\pi = 0$

$$0 \leq q \cdot p' \leq \frac{M_K^2}{2} .$$

Following the method which we established in the previous paragraph, we study

$$M_{\mu\nu} = \int dx\, e^{ip' \cdot x} \langle 0 | T(A_\mu^{\pi^0}(x)\, V_\nu^{K^-}(0)) | K^+(p) \rangle$$

and find

$$-ip'^{\mu} M_{\mu} = \int dx\, e^{ip' \cdot x}\, \delta(x_0)$$

$$\langle 0 | [A_0^{\pi^0}(x), V_\nu^{K^-}(0)] | k \rangle$$

$$= \frac{1}{\sqrt{2}} \langle 0 | A_\nu^{K^-} | K^+(p) \rangle$$

using the $SU_3 \times SU_3$ algebra of currents proposed earlier. The left-hand side of this equation has a very simple form in the limit $p' \to 0$:

$$-ip'^{\mu} M_{\mu\nu} \to -\, if_\pi \langle \pi^0(0) | V_\nu^{K^-} | K^+(p) \rangle .$$

Therefore

$$-if_\pi \langle \pi^0(0) | V_\nu^{K^-} | K^+(p) \rangle = \frac{1}{\sqrt{2}} \langle 0 | A_\nu^{K^-} | K^+ \rangle .$$

Now

$$\langle \pi^0(0) | V_\nu^{K^-} | K^+(p) \rangle = (f_+(0) + f_-(0))\, p_\nu,$$

$$\langle 0 | A_\nu^{K^-} | K^+(p) \rangle = f_K(-ip_\nu)$$

where f_K may be measured in the decay $K^+ \to e^+ + \nu$. This gives the simple relations

$$f_+(0) + f_-(0) = \frac{f_K}{f_\pi}$$

between the parameters of $K^+ \to \pi^0 e^+ \nu$ and of $K^+ \to e^+ \nu$.

How can this relation be compared with experiment? Let us imagine that our theory is such that the physical mass of the pion is a variable parameter. Then we must write

$$f_\pm = f_\pm(p' \cdot p, p'^2),$$

p'^2 being the pion mass. Let us further imagine that, at least over the range from zero to 135 mev, form factors depend very weakly on pion mass:

$$f_\pm(p' \cdot p, 0) \cong f_\pm(p' \cdot p, \mu_\pi^2).$$

In such circumstances we could write

$$f_+(0, \mu_\pi^2) + f_-(0, \mu_\pi^2) \cong \frac{f_K}{f_\pi} .$$

Unfortunately, $f_\pm(0, \mu_\pi^2)$ cannot be measured directly because the lower limit on the physical spectrum is

$$p' \cdot p = M_K \mu_\pi .$$

However, since $f_\pm$ are analytic functions of the variable $p' \cdot p$, we can analytically continue out of the physical region to discover the value of the form factors at $p' \cdot p = 0$. Therefore, if the statement that form factors depend only weakly on the value of the physical pion mass is correct, we can hope to compare low-energy theorems, such as the above equation, with experiment.

There is no a priori reason why this hypothesis should be correct, although it seems reasonable in view of the smallness of the pion mass on the scale of typical strong interaction energies. We believe now that it is basically correct since if we use it in conjunction with the current commutation relations, we can derive a large number of results, all of which are in excellent agreement with experiment. Since time is lacking, we cannot justify these statements by going into the applications in detail. The reader can convince himself by referring to some of the papers mentioned in the bibliography.

We shall study one final application of the methods expounded here, which is quite eloquent of their power. We shall consider the decay $K^+ \to \pi^+\pi^- e^+ \nu$, which, in spite of its small rate, is being studied in detail by several experimental groups. The matrix element is

$$M = \bar{u}_\nu \gamma^\mu (1 + \gamma_5)\, v_e \langle \pi^+(q)\pi^-(q') \,|(V_\mu^{K^-} + A_\mu^{K^-})|\, K^+(p)\rangle .$$

Lorentz invariance allows us to write

$$\langle \pi^+(q)\ \pi^-(q')\ |A_\mu^{K^-}|K^+(p)\ \rangle$$

$$= f_1(q+q') + f_2(q-q')_\mu + f_3(p-q-q')_\mu .$$

$$\langle \pi^+(q)\ \pi^-(q')\,|\,V_\mu^{K^-}|K^+(p)\rangle$$

$$= g\epsilon_{\mu\nu\lambda\sigma} q^\nu q'^\lambda p^\gamma,$$

each form factor being a function of $p \cdot q$, $p \cdot q'$ and $(q+q')^2$. In practice we can measure only f_1, f_2, and g. Because of the Dirac equation, f_3 is multiplied by m_e in the matrix element.

To obtain a useful theorem we study

$$M_{\mu\nu\lambda} = \int dxdy\ e^{i(q \cdot x + q' \cdot y)}$$

$$\langle 0\,|T(A_\mu^{\pi^-}(x)\,A_\nu^{\pi^+}(y)\,A_\lambda^{K^+}(0))\,|K^+(p)\rangle$$

By taking successive divergences, we find

$$(-iq^\mu)(-iq'^\nu)M_{\mu\nu\lambda} = \frac{i(q'-q)^\nu}{2} \int dy\ e^{i(q+q')\cdot y}$$

$$\langle 0\,|T(V_\nu^{\pi^0}(q)\,A_\lambda^{K^-}(0))|\rangle + \frac{1}{4}\langle 0\,|A_\lambda^{K^-}|K^+(p)\rangle$$

(using the equal-time commutators of currents several times). On the other hand, by picking out pole terms in $M_{\mu\nu\lambda}$ we see that

$$(-iq^{\mu})(-iq'^{\nu})M_{\mu\nu\lambda} = f_{\pi}^{2}\langle\pi^{+}(q)\pi^{-}(q')|A_{\lambda}^{K^{-}}|K^{+}(p)\rangle$$

+ (terms of 2nd. order in pion momenta).

Now define

$$M_{\nu\lambda}(l) = \int dy\, e^{il\cdot y}\langle 0|T(V_{\nu}^{\pi^{0}}(q)A_{\lambda}^{K^{-}}(0))|K^{+}\rangle$$

$$= 0\left(\frac{1}{l}\right) + 0(l^{0}) + 0(l') + \ldots \, .$$

We shall find it easy to identify the first two terms, in such an expansion in powers of l. The $0(^{1}/l)$ piece can only come from the pole term diagrammatically represented as

$$K^{+} \underset{p}{------} \overset{V_{\nu}}{x} \underset{p-l}{------} x\, A_{\lambda} = \frac{f_{K}}{2}\,\frac{(p-l)_{\lambda}(2p-l)_{\nu}}{l^{2}-2p\cdot l} \, .$$

The constant piece must have the form $A\,\delta_{\nu\lambda} + B\,p_{\nu}p_{\lambda}$. On the other hand, by using the current algebra

$$-il^{\nu}M_{\nu\lambda} = -\frac{1}{2}\langle 0|A_{\lambda}^{K^{-}}|K^{+}\rangle = \frac{if_{K}}{2}\,p_{\lambda}$$

$$= -iAl_{\lambda} - iBp\cdot l p_{\lambda} - \frac{if_{K}}{2}\,\frac{l\cdot(2p-l)(p-l)_{\lambda}}{l^{2}-2p\cdot l}$$

$$+ 0(l^{2})$$

$$= -iAl_\lambda - iBp \cdot l\, p_\lambda + \frac{if_K}{2} (p - l)_\lambda + 0(l^2).$$

These equations can be satisfied only if

$$B = 0 \quad A - \frac{f_K}{2} .$$

Then

$$M_{\nu\lambda} = - \frac{f_K}{2} \frac{(p - l)_\lambda (2p - l)_\nu}{2p \cdot l - l^2} - \frac{f_K}{2} \delta_{\nu\lambda} + 0(1),$$

and

$$f_\pi^2 \langle \pi^+(q)\, \pi^-(q') \,|\, A_\lambda^{K^-} \,|\, K^+\rangle = \frac{if_K}{4} \Big\{ (q + q')_\lambda$$

$$+ (q - q')_\lambda + (1 - \frac{p \cdot (q' - q)}{p \cdot (q' + q)}) (p - q - q')_\lambda \Big\}$$

$$+ 0(q^2).$$

This allows us to conclude that the measurable form factors are

$$f_1 = f_2 = \frac{f_K}{4} + \text{(terms of first order in pion momenta.)}$$

This prediction appears to agree quite well with the experiments. For example, the ratio f_1/f_2 averaged over the spectrum is measured to be .9 ± 1, which is to be compared with our prediction of 1. The g form factor probably makes a small contribution to the rate because of the many factors of momentum multiplying it. If this is so, the rate is determined by $f_{1,2}$ alone. The rate predicted by the above values of the g's agrees well with the measured decay rate.

There are many other applications of this scheme which give equally interesting results. Limitations on time forbid us to go into them in any detail, so we shall give a selected bibliography of applications, as well as the original references to the applications studied here.

BIBLIOGRAPHY

The current algebra was introduced into physics by Gell-Mann:

1. M. Gell-Mann Phys. Rev. 125 1967 (1962).

The idea of using hypotheses about the divergence of the axial-vector current was introduced in the context of dispersion theory by Goldberger and Treiman, refined by Gell-Mann and Levy, and extended by Nambu and co-workers and Adler. The paper by Adler is most relevant for the development given here.

2. Goldberger and Treiman Phys. Rev. 110 1178 (1958).
3. Gell-Mann and Levy, Nuovo Cimento 16, 705 (1960).
4. Nambu and Lurie, Phys. Rev. 125, 1429 (1961).
5. Adler. Phys. Rev. 139, B1638 (1965).

These two sets of ideas were combined independently by Adler and Weisberger to give a spectacularly successful sum rule for g_A, the axial vector renormalization constant.

6. Adler Phys. Rev. Letters 14, 1051 (1965).

7. Weisberger Phys. Rev. 143, 1302 (1966).
The high-energy neutrino sum rule given in the text was first derived by Adler (the derivation given in the text is due to the author).
8. Adler, Phys. Rev. 140, B736 (1965).
The applications of current-algebra techniques to weak processes can be classified roughly as follows (9 and 10 are the references for the leptonic decay chain studied in the text).
Leptonic Decays.
9. Callan and Treiman Phys. Rev. Letters 16, 153 (1966).
10. S. Weinberg Phys. Rev. Letters 17, 336 (1966).
Non-Leptonic Decays.
11. Suzuki Phys. Rev. Letters 15 986 (1966).
12. Callan and Treiman (see 9).
13. Hara and Nambu Phys. Rev. Letters 16, 875 (1966). Further elaboration of the current algebra scheme as a method of obtaining low-energy theorems for πN and $\pi\pi$ scattering lengths can be found in
14. S. Weinberg Phys. Rev. Letter 16, 879 (1966).
This list is of course not complete. It is intended only to give the interested reader an entrée into the extensive literature on the subject.

THE PHYSICS BEHIND ANALYTICITY IN MOMENTUM TRANSFER AND HIGH ENERGY BOUNDS

Ralph Roskies

THE PHYSICS BEHIND ANALYTICITY IN MOMENTUM TRANSFER AND HIGH ENERGY BOUNDS

Ralph Roskies
Yale University

INTRODUCTION

The ideas of dispersion relation and analytic S matrix theory have prodded physicists to study the analytic properties of scattering amplitudes for the past several years. In fact, such investigations go back more than twenty years to the early papers of Heisenberg on S matrix theory and to the work of Jost,[1] Hu[2] and others who studied the properties of the S matrix as an analytic function of the energy for fixed angular momentum, in non-relativistic quantum theory. The results were unfortunately more of mathematical and formal interest, because the physical basis of the analyticity was not understood.

The physics behind such analyticity began being appreciated with the work of Schutzer and Tiomno[3] and Van Kampen.[4] Following a suggestion of Kronig,[5] they exploited the idea of macroscopic causality to derive the desired analytic properties. Macroscopic causality meant that there could be no scattered wave until the incident wave had reached the target. Admittedly there are difficulties[6] with this criterion, since one cannot build wave packets with sharp wave fronts with particles of non-zero mass. But in some sense, one understood what lay behind analyticity in the energy. The next great step was taken by Gell-Mann, Goldberger and Thirring.[8] Working in a field theoretic framework, they translated macroscopic causality into microscopic causality (that is, local commutativity) to derive analytic properties of the relativistic S matrix. This became the basis of future attempts to derive such results in field theory, since local commutativity was incorporated into the axioms of field theory.[9] From then on, progress was very rapid, spearheaded by the work of Mandelstam, Lehmann, Regge and many others, and it has become fashionable to discuss analyticity in angular momentum and momentum transfer as well as in energy.

In these lectures, I will concentrate on analyticity in momentum transfer t for fixed energy. It had long been clear that the analytic domain in t was closely related to the range of the forces between the scattering particles. To make this plausible, consider the elastic scattering of two particles with momentum k, total energy $\sqrt{s}$ and scattering angle θ in the center of mass system. Then

$$t = -2k^2(1 - \cos\theta). \qquad (1)$$

For fixed s, k is fixed and t is linearly related to $\cos\theta$. The simplest Feynman graph

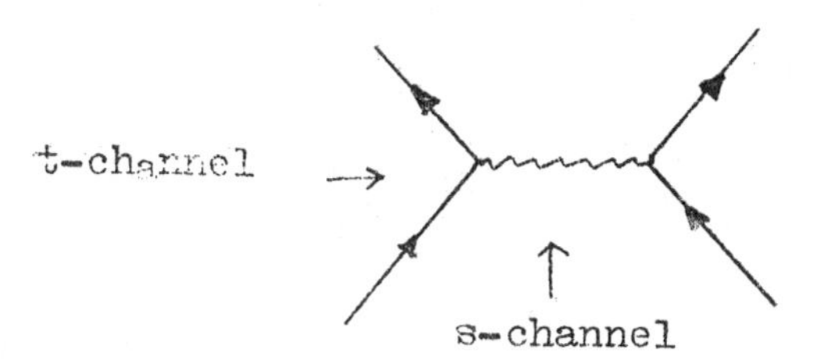

gives rise to a pole in the amplitude at $t = m_e^2$, where m_e is the mass of the exchanged particle. The singularity closest to $t = 0$ is then determined by the lightest mass which can be exchanged. On the other hand, the range of the forces is inversely proportional to the lightest mass which can be exchanged, so that the singularity closest to $t = 0$ is determined by the range of the forces.

These considerations were made much more precise in a paper by Omnes[10] which shed considerable light on the basis of t analyticity. Before presenting his ideas, let me briefly review what is known about t analyticity in field theory.

Lehmann[11] proved that the elastic amplitude for 2 particle scattering at fixed s is analytic in $z = \cos\theta$ in an ellipse with foci at $z = \pm 1$ and with semi major axis

$$z_0(s) = \left[1 + \frac{(M_1^2 - m_1^2)\,(M_2^2 - m_2^2)}{k^2(s - (m_1 - m_2)^2)}\right]^{1/2} \quad (2)$$

where m_i is the mass of particle i, and M_i is the lowest invariant mass of all many-particle states with the same quantum numbers as those of particle i. In the t plane, this means analyticity in an ellipse with foci at $t = 0$, $t = -4k^2$ and apex

$$t_0 = -2k^2(1 - z_0) \ . \tag{3}$$

For large s, the apex approaches $t = 0$, so that there is no neighbourhood of $t = 0$ which is within the analytic domain for all energies.

Greenberg and Low[12] showed that Lehmann's result implies that the total cross section $\sigma(s)$ satisfies

$$\sigma(s) < C \, s \ln^2 s \tag{4}$$

for large s, where C is a constant. They also pointed out that if one could establish analyticity in a neighbourhood of $t = 0$ for all energies, then $\sigma(s)$ would satisfy

$$\sigma(s) < C \ln^2 s \tag{5}$$

for large s. The bound in (4) is known as the Greenberg-Low bound, whereas the bound in (5) is called the Froissart bound, and was established by Froissart[13] assuming the full Mandelstam representation. It is interesting to note that the observed cross sections seem to approach constants at high energies, and since $\ln^2 s$ is a slowly-varying function, the theoretical bound (5) is almost saturated.

Recently, Martin[14] was able to show that where one could establish dispersion relations in the energy, one could also establish a neighbourhood of analyticity around t = 0 independent of energy. Thus for $\pi\pi$ scattering, or for the more interesting πN case, the Froissart bound is a result of local field theory. Because one cannot establish the necessary energy dispersion relations to apply Martin's techniques, the best result on NN scattering is still the Greenberg-Low bound.

The proofs establishing these results suffer from some aesthetic and practical drawbacks. First, the underlying physics is obscured by the early introduction of the techniques of analytic continuation. In contrast, I shall present proofs of the results of Lehmann and Greenberg and Low, in which the physical basis is always clear. Analytic continuation is introduced only at the end of the discussion.

Secondly, the existing proofs suggest that the various bounds can be established only after proving analyticity. In the method to be discussed, they can be obtained directly, and we shall see that there are cases where the bounds are essentially valid, but there is no analytic domain for the scattering amplitude. I mention this to point out that often too much emphasis is placed on analyticity. To the extent that its physical consequences can be obtained independently, analyticity loses its relevance for physics.

Thirdly, the usual proofs assume that to each stable particle there corresponds an interpolating field satisfying local commutativity. There is no rigorous proof that this is true for composite particles if it is true for the elementary ones. Thus for scattering processes involving the deuteron, for instance, the Greenberg-Low bound has not been established. In view of the success of symmetry schemes which treat particles and resonances on an equal footing, we shall adopt a formulation of field theory in which particles are represented by interpolating fields built out of finite products of some basic fields at different space-time points.

For the moment, let me outline Omnes' simple picture and some improvements due to Kugler and Roskies[15] within potential theory. Then we shall apply the same physical ideas to the field theoretic case.

POTENTIAL THEORY

If one wants to relate t analyticity to the short range nature of the force, one must first decide how to characterize such a force. Omnes' idea is simple. Consider the scattering of two wave packets with average momentum $\bar{k}$ and $-\bar{k}$, and with impact parameter a. That is, if there were no interaction, the closest approach of the centers of the wave packets would be a.

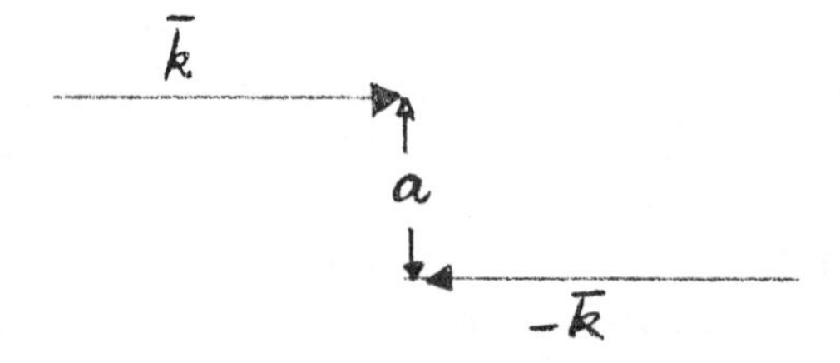

(We must have

$$\bar{k} \cdot \bar{a} = 0) . \tag{6}$$

If the forces are short range, one would expect the

norm of the scattered wave to fall off rapidly with increasing a. To see what sort of fall off to expect, let us turn to potential scattering. Since we certainly want a Yukawa potential to be an example of a short range force, let us study how fast the norm of the scattered wave falls with impact parameter.

Let φ be the free incident wave packet at time $t = 0$, S the S matrix, H the full Hamiltonian, H_0 the free Hamiltonian. Then

$$S\varphi = \lim_{\substack{t \to \infty \\ t' \to -\infty}} e^{iH_0 t} e^{-iH(t-t')} e^{-iH_0 t'} \varphi \tag{7}$$

$$= \lim_{\substack{t \to \infty \\ t' \to -\infty}} U(t,t') \varphi \tag{8}$$

$$(S - 1)\varphi = - \lim_{t \to \infty} \int_{-\infty}^{t} \frac{d}{dt'} U(t,t')\varphi \, dt' \tag{9}$$

$$\| (S - 1) \varphi \| \leq \lim_{t \to \infty} \int_{-\infty}^{t} \| \frac{d}{dt'} U(t,t') \varphi \| \, dt'$$

$$\leq \lim_{t \to \infty} \int_{-\infty}^{t} \| e^{iH_0 t} e^{-iH(t-t')} i(H - H_0) e^{-iH_0 t'} \varphi \| \, dt'$$

$$\leq \lim_{t \to \infty} \int_{-\infty}^{t} \| V e^{-iH_0 t'} \varphi \| \, dt' \tag{10}$$

using the unitarity of e^{iH_0t}, $e^{-iH(t-t')}$ and the definition of the potential

$$V = H - H_0 \, . \tag{11}$$

Thus we have

$$\| (S - 1) \varphi \| \leq \int_{-\infty}^{\infty} \| V\varphi(t') \| \, dt' \tag{12}$$

where $\varphi(t')$ is the incident wave packet propagating in the absence of interaction.

If we write

$$S = 1 + 2i\,T \tag{13}$$

then the scattered wave is $2i\,T\,\varphi$, and (12) gives

$$\|2T\varphi\| \leq \int_{-\infty}^{\infty} dt' \, \| V \varphi(t') \| \, . \tag{14}$$

This is an interesting inequality since it puts a bound on the T matrix which is linear in the potential.

We choose the incident packet to represent the physical situation outlined before - two packets with average momentum $\bar{k}$ and $-\bar{k}$ and impact parameter $\bar{a}$. Factoring out the center of mass motion, the normalized relative packet can be chosen to be

$$\varphi(\bar{p}, t) = \left[\frac{b}{\sqrt{\pi}}\right]^{3/2} e^{-(\bar{p} - \bar{k})^2 b^2/2} \, e^{-i\bar{p}\cdot\bar{a}} \, e^{\frac{-ip^2 t}{2m}} \, , \tag{15}$$

a Gaussian with average momentum $\bar{k}$ displaced from the origin by $\bar{a}$ at $t = 0$. The explicit form of the packet is not important; the Gaussian is chosen for convenience. At the end, of course, all reference to the wave packet will disappear since we are seeking information on the T matrix. The wave packet is merely a convenient tool to enable us to elicit some information about T. In x space

$$|\varphi(\bar{x}, t)|^2 = \left(\frac{1}{\sqrt{\pi}\, b(t)}\right)^3 e^{-(\bar{x} - \bar{a} - \bar{k}t/m)^2/b(t)^2} \tag{16}$$

with

$$b(t)^2 = b^2 + t^2/m^2 b^2 . \tag{17}$$

We see that φ represents a packet with width $b(t)$ whose center moves along the trajectory

$$\bar{x} = \bar{a} + \bar{k}t/m \quad . \tag{18}$$

b is the width of the packet at $t = 0$, and the linear dependence of $b(t)$ on t for large t is a manifestation of the spreading of the wave packet. The rate of spreading varies inversely as the initial width b.

To estimate the right side of (14), recall that

$$\|V \varphi (t) \|^2 = \int |V(x) \varphi(x, t)|^2 \, d^3x . \tag{19}$$

If

$$V(x) = C\, e^{-\mu x}/x \tag{20}$$

then for finite time, it is easy to see that the right side of (19) decreases exponentially with a. But at large times,

$$|\varphi(x,t)|^2 \to \left(\frac{mb}{\sqrt{\pi}\,t}\right)^3 e^{-k^2 b^2} \tag{21}$$

independent of a. Thus for large t, $\|V\varphi(t)\|$ is essentially independent of a, and so the right side of (14) does not fall off rapidly with a. This arises from the spreading of the wave packet. For large t, the packet has spread so much that it has forgotten about the initial displacement a. Thus, it does not follow that $\|T\varphi\|$ falls off exponentially with a.

But, from (21), we see that if we make b^2 proportional to a, then even at large times $|\varphi|^2$ will decrease exponentially with a. Thus to get exponential falloff of $\|T\varphi\|$ with a, we must choose the width b to increase with a according to

$$b^2 = \lambda a . \tag{22}$$

As pointed out above, increasing b decreases the effects of the spreading. On the other hand, since b only varies as $\sqrt{a}$, it is still true that the bulk of the incident packet does not overlap the scattering center.

To sum up, if one only increases the impact parameter leaving the width fixed, $\|T\varphi\|$ decrease like 1/a. But if we also make the width proportional to $\sqrt{a}$, $\|T\varphi\|$ falls off exponentially with a. A detailed proof of these statements is found in the Appendix.

With the packet

$$\varphi(\bar{p}, t = 0) = (\frac{\lambda a}{\pi})^{3/4} e^{-(\bar{p} - \bar{k})^2 \lambda a/2} e^{-i\bar{p} \cdot \bar{a}} \tag{23}$$

and with $$\lambda = 1/k \tag{24}$$

for convenience, (see Appendix), then for a Yukawa potential

$$\|T\varphi\| < C e^{-\tau a} \tag{25}$$

where C is some polynomially bounded function of a, k, and τ depends on μ and k, with

$$\tau \to \mu \tag{26}$$

for large k.

We then define a short range force as one in which (25) is satisfied, with φ given by (23) and τ some function of k approaching a finite non-zero limit as $k \to \infty$. It is now our task to discover what (25) implies about the T matrix.

Here we depart from Omnes' analysis, and follow that of Kugler and Roskies. (25) is an inequality on T for large impact parameter. But large impact parameter corresponds to a semiclassical situation. If k is the average momentum, a the impact parameter, then the incident beam has its angular momenta concentrated near $l_0 \equiv ka$. If we make a partial wave decomposition of φ, only those components with $l \approx l_0$ will be large. Consequently

$$\|T\varphi\| \approx |T_{l_0}(k)\, \varphi_{l_0}(k)| \tag{27}$$

and (25) becomes

$$|T_{l_0}(k)\, \varphi_{l_0}(k)| < C\, e^{-\tau a} = C\, e^{-\tau l_0/k}. \qquad (28)$$

But $\varphi_{l_0}(k)$, the component of φ with angular momentum $l_0 = ka$ should be large so that

$$|T_{l_0}(k)| < C\, e^{-\tau l_0/k} \qquad (29)$$

and therefore $T_l(k)$ decreases exponentially with l/k.

To make this more precise, recall that the unitarity of S implies

$$\frac{T - T^{\dagger}}{2i} \equiv \mathrm{Im}\, T = T^{\dagger} T \;. \qquad (30)$$

Thus

$$\|T\varphi\|^2 = (T\varphi,\, T\varphi) = (\varphi,\, T^{\dagger} T\varphi) = (\varphi,\, \mathrm{Im}\, T\, \varphi). \qquad (31)$$

If we write

$$\varphi(\bar{p}) = \sum_{lm} \varphi_{lm}(p, \bar{k}, \bar{a})\; Y_{lm}(\hat{p}) \qquad (32)$$

then

$$\|T\varphi\|^2 = \sum_{l} \int p^2\, dp \sum_{m=-l}^{l} |\varphi_{lm}(p, \bar{k}, \bar{a})|^2\, \mathrm{Im}\, T_l(p)$$
$$< C\, e^{-2\tau a} \;. \qquad (33)$$

By unitarity, $\mathrm{Im}\, T_l(p) > 0$, so that each term in the sum in (33) is positive and the inequality holds a fortiori if we replace the sum over l by a single term. Moreover we can restrict the integration to a small region around $p = k$. Assuming Im $T_l(p)$ is smooth in p, we can take it out of the integral and replace it by Im $T_l(k)$. By explicit calculation,[15] we can then evaluate the remainder, and we find, neglecting l independent factors of powers of a and k,

$$\text{Im } T_l(k) \frac{e^{-2ka}(2ka)^{2l}}{(2l)!} < C e^{-2\tau a} \quad . \tag{34}$$

The value of a which maximizes the coefficient of Im $T_l(k)$ is $a = l/k$ which confirms the discussion given earlier. Recall that $T_l(k)$ has no a dependence so that we can choose a at will. For large l, and $a = l/k$, we find

$$\text{Im } T_l(k) < C e^{-2\tau l/k} \tag{35}$$

which is the desired result - a bound on Im T_l for large l. Notice that all reference to the wave packet has disappeared, and we have an inequality on the T matrix. (We have chosen the wave packet to probe some aspect of the dynamics. By choosing $a = l/k$, we have picked out the lth partial wave of T; by varying a we get conditions on the different partial waves). If we had chosen a very different from l/k in (34), the coefficient of Im $T_l(k)$ would have been small, and no useful dynamical information about Im $T_l(k)$ would have been obtained.

What are some implications of (35)? Recall that

$$\text{Im } T(k, z) = \sum_l (2l + 1) \text{ Im } T_l(k) \, P_l(z) \; . \tag{36}$$

Now we invoke the following theorem:[16] The series

$$\sum_l a_l P_l(z)$$

converges and defines an analytic function of z in an ellipse with foci at $z = \pm 1$ and semi-major axis

$$z_0 = \frac{1}{2} (r_0 + r_0^{-1}) \tag{37}$$

where r_0 is the radius of convergence of the power series

$$\sum a_l x^l,$$

provided that $r_0 \geq 1$. It follows then from (35) that Im T (k, z) is analytic in z in an ellipse with foci at $z = \pm 1$ and semi-major axis

$$z_0 = \frac{1}{2} (e^{2\tau/k} + e^{-2\tau/k}) \; . \tag{38}$$

For large k, $\tau \approx \mu$,

$$z_0 \approx 1 + 2\mu^2/k^2 \; . \tag{39}$$

In the t plane, Im T is analytic in an ellipse with foci at $t = 0$, $t = -4k^2$ and for large k, the apex is at $4\mu^2$. This is exactly the region found by Martin for $\pi\pi$

scattering with $\mu = m_\pi$ (corresponding to a long range potential $C\ e^{-\mu r}/r$). Thus (25), the statement of short range forces, has led to the analytic domain established by Martin in relativistic field theory.

Notice that until the very end, all the discussion was about physical quantities. The bound on the partial waves led to the analyticity and not vice-versa as in the usual proofs.[17]

The proof can also be reversed. From the analytic domain of Im T, one can deduce a bound like (35). This then leads to a bound like (25) so the short range condition is equivalent to the analyticity.

By using unitarity

$$\text{Im } T_l \geq |T_l|^2 \tag{40}$$

one can show that T(k, z) is analytic in an ellipse with foci at $z = \pm 1$, and semi-major axis

$$\frac{1}{2}\left(e^{\tau/k} + e^{-\tau/k}\right) .$$

For large energy, this means analyticity in t in an ellipse with apex at $t = \mu^2$. Analyticity in a larger ellipse would exclude the possibility of a pole at $t = \mu^2$ and is therefore undesirable.

The next consequence of (35) is the Froissart bound. With our normalizations

$$\sigma = \frac{4\pi}{k^2} \sum_l (2l + 1) \text{ Im } T_l (k) . \tag{41}$$

Moreover, unitarity implies that

$$0 \leq \mathrm{Im}\, T_l\,(k) \leq 1\,. \tag{42}$$

For large l, from (35), restoring the possible k dependence of C

$$\mathrm{Im}\, T_l(k) < C_1 k^N e^{-2\tau l/k} \tag{43}$$

where C_1 is some constant. This shows that for large l, Im T_l falls off rapidly, so that the contribution to σ will be appreciable only for $l \leq l_0$, with l_0 such that the right side of (43) is of order 1. This means

$$l_0 \approx C_2 k \ln k \tag{44}$$

where the ln k factor is chosen to compensate for the factor k^N in (43). One easily sees that the contribution to σ from $l \geq l_0$ is bounded by terms of order ln k/k which vanish as $k \to \infty$. For $l < l_0$, we replace Im T_l (k) by its upper bound according to (42) and find

$$\begin{aligned} \sigma &\leq \frac{4\pi}{k^2} \sum_{l=0}^{l_0} (2l+1) \\ &\leq 4\pi\, C_2^2 \ln^2 k \end{aligned} \tag{45}$$

which is the Froissart bound.

(In potential scattering, one can keep track of the powers of a and k disregarded in (35). One can then show that $\sigma \to 0$ instead of blowing up like $\ln^2 k$).

At this point, I wish to emphasize that in the derivation of the Froissart bound, (43) was used only to determine the value of l_0, i.e. the cut off value of l. It was not important that the falloff was exponential, but rather that it was rapid in l/k. If we had had

$$\mathrm{Im}\, T_l(k) < C\, k^N\, e^{-\tau_1 \sqrt{l/k}} \tag{46}$$

then the only difference would have been

$$l_0 \approx k \ln^2 k \tag{47}$$

instead of (44), so that

$$\sigma < C \ln^4 k \quad . \tag{48}$$

In general, if $\mathrm{Im}\, T_l$ were bounded by a function falling off faster than any power of l/k, we could obtain

$$\sigma < C\, k^{\epsilon} \qquad \epsilon > 0, \text{ arbitrarily small.} \tag{49}$$

It is interesting to note that bounds like (46) allow for singularities in Im T(k, z) at the physical point $z = 1$. The partial wave expansion will converge only in the physical region, z real $|z| < 1$. Thus there is no analyticity, but the bound on σ is hardly altered. This shows that these bounds are really independent of analyticity, and are more closely related to a rapid falloff in l/k than to a strictly exponential one. Moreover, if the potential had been

$$V = C\, e^{-\mu \sqrt{r}},$$

then the analysis we have presented would give rise to (46) and so we see that it is easy to construct examples of potentials for which (48) holds, but which have no analytic domains for T. In our method of dealing directly with the partial wave amplitude, we can establish the bound directly, without any considerations of analyticity.

Suppose finally that the potential had been energy dependent

$$V = C\, e^{-\mu r/k},$$

with a range increasing linearly with k. We could then show that

$$\|T\varphi\| < C\, e^{-\tau a/k} \tag{50}$$

which would imply

$$\mathrm{Im}\, T_l\, (k) > C\, e^{-\tau l/k^2} . \tag{51}$$

Again, we would have analyticity in an ellipse in the t plane, but with the apex approaching $t = 0$ like $1/k^2$ for large k. The bound on the cross section would be

$$\sigma < C\, k^2 \ln^2 k \tag{52}$$

because the cutoff value l_0 would be of order $k^2 \ln k$. This analytic domain and the bound on σ are precisely those found by Lehmann and by Greenberg and Low. They are to be interpreted as arising from a potential whose range increases linearly with k. As before, if the potential were

$$C\, e^{-\mu \sqrt{r/k}},$$

(52) would be essentially unchanged, but there would be no domain of analyticity, so that the Greenberg-Low bound is not dependent on analyticity in the Lehmann ellipse.

FIELD THEORY

In the next lectures, I would like to examine some consequences of the physical ideas outlined previously for relativistic field theory, not only for their intrinsic interest, but also to illustrate that the Haag-Ruelle[18, 19] formulation of scattering theory is a natural extension of what we are used to in potential scattering.

There are, of course, many differences, both formal and physical. The mathematical techniques of field theory are more complicated and we must borrow results from Hilbert space theory and the theory of distributions. More importantly, we do not have the equations of motion in field theory, so that we must postulate various general properties of the fields and hope that these allow us to deduce some useful results. Time does not permit me to give a detailed account of the field theoretic tools[20] which are necessary for the discussion, but let me outline their basis briefly.

In order to study the scattering problem, we must first learn how to describe the particles occurring in the theory (the one particle problem) and then see how to construct scattering states composed of several particles (in our case, the two particle problem). Then we can define the S matrix in terms of these states.

Consider the elastic scattering of two spinless particles X_1 and X_2 of mass m_1 and m_2.

I. To particle X_i we assign a creation operator $B_i^\dagger$, and assume that if $|0>$ is the vacuum state, $B_i^\dagger |0>$ is a state with the quantum numbers of X_i. Moreover, we suppose that $B_i^\dagger$ can be written as

$$B_i^\dagger = \int d^4x_1 \ldots d^4x_n \phi_i(x_1, x_2 \ldots x_n) A(x_1) \ldots A(x_n) \tag{53}$$

where the A's are local field operators,[21] i.e.

$$[A(x_1), A(x_2)] = 0 \qquad (x_1 - x_2)^2 < 0 \tag{54}$$

and ϕ is a C^∞ function which falls off faster than any inverse power of the Euclidean norm

$$\|\underline{x}\| = (\sum_i \|x_i\|^2)^{1/2} \tag{55}$$

with

$$\|x_i\| = (x_{i0}^2 + \vec{x}_i^2)^{1/2}. \tag{56}$$

The rapid decrease of ϕ assures that $B_i^\dagger$ is somewhat localized around the origin.

II. We postulate the existence of a unitary representation $U(a, \Lambda)$ of the Poincaré group (inhomogeneous Lorentz group) satisfying

$$U(a_1, \Lambda_1)\, U(a_2, \Lambda_2) = U(a_1 + \Lambda_1 a_2, \Lambda_1 \Lambda_2) \tag{57}$$

where a_1, a_2 represent space-time translations, and Λ_1, Λ_2 represent homogeneous Lorentz transformations. Moreover, the vacuum is invariant under the group

$$U(a, \Lambda)\, | 0 \rangle = | 0 \rangle \ . \tag{58}$$

The fields $A(x)$ are assumed to transform according to

$$U(a, \Lambda)\, A(x)\, U(a, \Lambda)^{-1} = A(\Lambda x + a). \tag{59}$$

We denote $U(a, 1)$ by $U(a)$. Then we define

$$B_i^{\dagger}(x) = U(x)\, B_i^{\dagger}\, U(x)^{-1}. \tag{60}$$

The field $B_i^{\dagger}(x)$ is roughly localized around x. It is noteworthy that $B_i(x)$ does not satisfy local commutativity i.e. (54), and so the interpolating field for particle i is not a local field.

By Stone's theorem,[22] we can write

$$U(x) = e^{iP \cdot x} \tag{61}$$

with

$$P \cdot x = P_0 x_0 - \overline{P} \cdot \overline{x} \ . \tag{62}$$

P_μ are a set of four commuting self-adjoint operators which we shall interpret as the energy-momentum operators.

III. The spectrum of P corresponds to the momenta of physical particles. The vacuum is a normalizable state with $P_\mu = 0$, the single particle states have

$$P^2 = m_i^2 \qquad (P^2 = P_0^2 - \overline{P}^2)$$

and the many particle states have

$$P^2 > (\text{sum of all masses})^2 .$$

It will be assumed that there are no zero mass particles in theory, which corresponds to the assumption of short range forces in potential theory. The spectrum of P is thus

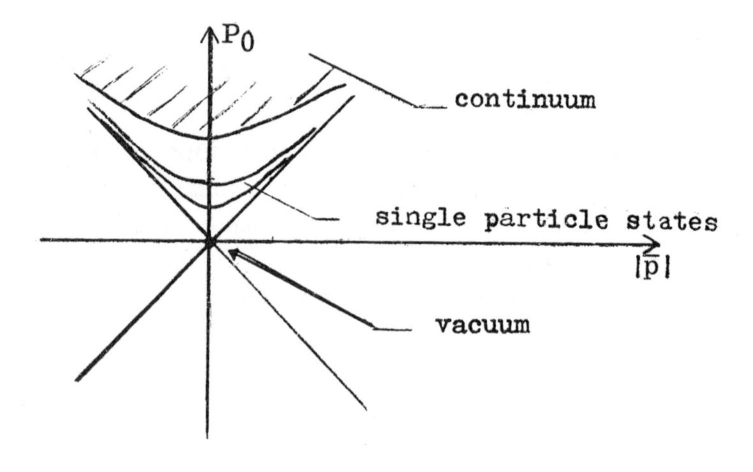

Let $|p, i>$ denote a non-normalizable state of momentum p with the quantum numbers of particle X_i. Then

$$< p, i \mid B_i^\dagger (x) \mid 0 > = < p, i \mid U(x)\, B_i^\dagger\, U(x)^{-1} \mid 0 >$$

$$= < p, i \mid e^{iP \cdot x}\, B_i^\dagger \mid 0 >$$

$$= e^{ip \cdot x} < p, i \mid B_i^\dagger \mid 0 >$$

$$= e^{ip \cdot x}\, g_i(p) \qquad (63)$$

where $g_i(p)$ depends on the test function $\phi_i(x_1, \ldots x_n)$. One can choose ϕ_i so that its Fourier transform

$$\tilde{\phi}_i (p_1, \ldots p_n)$$

satisfies

$$\tilde{\phi}_i(p_1, \ldots p_n) = 0 \text{ if } \left(\sum_{j=1}^{n} p_j \right)^2 > M_i^2 \qquad (64)$$

with M_i the lowest mass in the continuum with the quantum numbers of X_i. Then $g_i(p)$ vanishes for $p^2 > M_i^2$, and therefore $B_i^\dagger (x) \mid 0 >$ is a one-particle state only, because its components in the continuum vanish.

To create a normalizable one-particle state, let $f_i(x)$ be a normalizable solution of the Klein-Gordon equation of mass m_i

$$(\Box + m_i^2)\, f_i(x) = 0 . \tag{65}$$

Then

$$f_i(x) = \frac{1}{(2\pi)^3} \int \tilde{f}_i(\bar{p})\, e^{-ip\cdot x} \frac{d^3p}{2p_0} \tag{66}$$

$$p_0^2 - \bar{p}^2 = m_i^2$$

$$p_0 > 0 .$$

Let us define

$$B_{f_i}^{\dagger}(t) = i \int B_i(x)^{\dagger} \overleftrightarrow{\partial}_0 f_i(x)\, d^3x . \tag{67}$$

Then $B_{f_i}^{\dagger}(t)$ creates a one-particle state with wave function

$$\begin{aligned} h_i(p) &= \langle p, i \,|\, B_{f_i}^{\dagger}(t) \,|\, 0 \rangle \\ &= i \int d^3x \langle p, i \,|\, B_i^{\dagger}(x) \,|\, 0 \rangle \overleftrightarrow{\partial}_0 f_i(x) = g_i(\bar{p})\, \tilde{f}_i(\bar{p}) \end{aligned} \tag{68}$$

where $g_i(\bar{p})$ is the restriction of $g_i(p)$ to the mass shell of particle i.
Moreover, from (68),

$$\frac{d}{dt} B_{f_i}^{\dagger} \,|\, 0 \rangle = 0 \tag{69}$$

so that the one-particle state

$$B_{f_i}^{\dagger}(t) \mid 0 >$$

is time independent.

With our assumptions, Ruelle[19] has shown that

$$\lim_{t \to \pm\infty} B_{f_1}^{\dagger}(t) \; B_{f_2}^{\dagger}(t) \mid 0 > \tag{70}$$

exists in the strong topology, and one can define asymptotic states

$$\lim_{t \to \pm\infty} B_{f_1}^{\dagger}(t) \; B_{f_2}^{\dagger}(t) \mid 0 > \; = B_{f_1}^{\substack{out \dagger \\ in}} \; B_{f_2}^{\substack{out \dagger \\ in}} \mid 0 > \tag{71}$$

where $B^{\substack{in \\ out}}$ is an object behaving like a free field. This means that at large times, the state

$$B_{f_1}^{\dagger}(t) \; B_{f_2}^{\dagger}(t) \mid 0 > \tag{72}$$

behaves like a state of two independent particles with wave packet $h_1(p_1)h_2(p_2)$. Recall that the B's are interacting fields, and at finite times (72) is much more complicated than a two-particle state. Ruelle's proof is valid for an arbitrary number of particles, and allows us to define asymptotic in and out states to define the S matrix. It is important for Ruelle's proof that $B_{f_i}^{\dagger}(t)$ create only one particle states with normalizable wave packets.

The S matrix for the elastic scattering of

particles X_1, X_2 with wave packets h_1, h_2 into packets h_3, h_4 is

$$<h_3 h_4^{out} | h_1 h_2^{in}> = <h_3 h_4^{in} | S | h_1 h_2^{in}>. \qquad (73)$$

Ruelle's convergence condition allows us to write this matrix element as a limit of the vacuum expectation values of certain field operators.

We choose however to rewrite this in a form very similar to the potential theory case. Writing

$$S = 1 + 2i\, T \qquad (74)$$

$$S^\dagger \, |h_1 h_2^{in}> = |h_1 h_2^{out}> \qquad (75)$$

$$-2\, iT^\dagger \, | h_1 h_2^{in}> = | h_1 h_2^{out}> - |h_1 h_2^{in}>$$

$$= \int_{-\infty}^{\infty} \frac{d}{dt} B^\dagger_{f_1}(t)\, B^\dagger_{f_2}(t) \, | 0 > dt \qquad (76)$$

$$< h_1 h_2^{in} | T\, T^\dagger | h_1 h_2^{in}> = \| T^\dagger |h_1 h_2^{in}> \|^2$$

$$= \| \frac{1}{2} \int_{-\infty}^{\infty} \frac{d}{dt} B^\dagger_{f_1}(t)\, B^\dagger_{f_2}(t) \, | 0 > dt \, \|^2$$

$$= \| \frac{1}{2} \int_{-\infty}^{\infty} \dot{B}_{f_1}^{\dagger}(t) \; B_{f_2}^{\dagger}(t) \; |0> dt \; \|^2$$

$$\leqq \left(\frac{1}{2} \int_{-\infty}^{\infty} \| \dot{B}_{f_1}^{\dagger}(t) \; B_{f_2}^{\dagger}(t) \; |0> \| \; dt \right)^2 \qquad (77)$$

where we have defined

$$\dot{B}_f^{\dagger}(t) = \frac{d}{dt} B_f^{\dagger}(t) \qquad (78)$$

and we have used (69). Expanding (77) using (67)

$$\| T^{\dagger} |h_1 h_2{}^{in}> \| \leq \frac{1}{2} \int_{-\infty}^{\infty} dt$$

$$\| \int d^3x_1 d^3x_2 \, (\Box + m_1^2) \, B_1^{\dagger}(x_1) \; B_2^{\dagger}(x_2) \, |0>$$

$$\overleftrightarrow{\partial}_{02} f_1(x_1) f_2(x_2) \; \| \, . \qquad (79)$$

This looks very much like (14) with f_1 f_2 as the wave packet (the time dependence is understood and $x_{10} = x_{20} = t$) and with

$$\| (\Box + m_1{}^2) \; B_1^{\dagger}(x_1) \; B_2^{\dagger}(x_2) \; | \; 0> \|$$

as the potential.

So far the discussion has been very general, and we have merely set up the formalism. Now we must exploit our physical picture of scattering at large impact parameters. There are two question to investigate—the construction of appropriate wave packets, and the short range character of the "potential".

The choice of wave packets is up to us; all reference to them will disappear at the end. To describe the scattering of two particles with average center of mass momentum $\bar{k}$, impact parameter $\bar{a}$, let

$$\tilde{f}_i\,(p) = \exp\left(\frac{1}{2}b_i^2\,(p - k_i)^2 + ip \cdot a_i\right) \tag{80}$$

where

p is a four-vector of mass m_i

k_i is the average four-momentum, also of mass m_i

a_i the displacement, chosen so that

$$\bar{k}_i \cdot \bar{a}_i = 0 \tag{81}$$

and

$$a_{i_O} = 0. \tag{82}$$

Since p and k are four-vectors of mass m_i, $(p - k)^2 < 0$ and f_i peaks at $p = k$ so that k is the average momentum. By direct computation, or using the method of stationary phase, $f_i\,(x)$ has its maximum at

$$\bar{x} = \bar{a}_i + \bar{v}_i\,t \qquad \bar{v}_i = \bar{k}_i/k_{iO}\,. \tag{83}$$

In the center of mass system, we shall take

$$\bar{k}_1 = -\bar{k}_2 \equiv \bar{k} \qquad (84)$$

and $\bar{a}_1$ and $\bar{a}_2$ in opposite directions so that the impact parameter will be $a = |\bar{a}_1 - \bar{a}_2|$.

However, as has already been pointed out, the effective wave packet is not $f(\bar{p})$ but $h(p)$ defined in (68). If $g(\bar{p})$ is a function of $\bar{p}^2$, such that $g = 1$ around $\bar{p} = 0$ and decreases away from $\bar{p} = 0$, then $g(\bar{p})\,\tilde{f}(\bar{p})$ will not necessarily peak at $p = k$ and the character of the effective packet will not be that of $\tilde{f}(\bar{p})$. To rectify this, we can redefine $g(p)$ to be a function of k as well. If Λ_i is the Lorentz transformation taking

$$\begin{pmatrix} m_i \\ o \end{pmatrix} \rightarrow \begin{pmatrix} k_{io} \\ \bar{k}_i \end{pmatrix} \qquad (85)$$

we use the function

$$\phi_i^{\Lambda_i}(x_1, x_2, \ldots x_n) \equiv \phi_i(\Lambda_i^{-1} x_1, \ldots, \Lambda_i^{-1} x_n) \qquad (86)$$

rather than $\phi_i(x_1, \ldots x_n)$ in the definition of B_i (see (53)).

Then

$$B_i^{\Lambda_i +} = \int d^4x_1 \ldots d^4x_n\, \phi_i^{\Lambda_i}(x_1 \ldots x_n)\, A(x_1) \ldots A(x_n)$$

$$= \int d^4x_1 \ldots d^4x_n\, \phi_i(x_1 \ldots x_n)$$

$$A(\Lambda_i x_1) \ldots . A(\Lambda_i x_n)$$

$$= U(\Lambda_i) \cdot B_i^{\dagger}\, U(\Lambda_i)^{-1} \qquad (87)$$

so that, as in (63),

$$\langle p,\, i\,|B_i^{\Lambda_i^{\dagger}}(x)\,|0\rangle = e^{ip\cdot x} \langle p,\, i\,|B_i^{\Lambda_i^{\dagger}}(0)\,|0\rangle$$

$$= e^{ip\cdot x} \langle p,\, i\,|U(\Lambda_i)\, B_i^{\dagger}(0)\,|0\rangle$$

$$= e^{ip\cdot x} \langle \Lambda_i^{-1}\, p,\, i|\, B_i^{\dagger}(0)\,|0\rangle$$

$$= e^{ip\cdot x}\, g_i(\Lambda_i^{-1}\, p). \qquad (88)$$

Comparing (88) with (63), the only change involved in using $B^{\Lambda}i$ instead of B is that

$$g_i(p) \rightarrow g_i(\Lambda_i^{-1}\, p)$$

which is always peaked around $\overline{p} = \overline{k}_i$. We shall interpret all B's in equation (77) - (79) to be $B^{\Lambda}i$'s. The result is to make the effective wave packet

$$\tilde{f}(p)\; g\,(\Lambda^{-1}\, p)$$

which is always peaked at $\overline{p} = \overline{k}$. I shall indicate later what would have happened if we hadn't made the Lorentz transformation.

As in the potential case, the width b_i^2 will be made proportional to a, to minimize the effects of spreading. So much for the wave packets.

As already mentioned

$$V^k(x_1, x_2 \equiv \| (\Box + m_1^2) B_1^{\Lambda_i^\dagger}(x_1)$$

$$B_2^{\Lambda_2^\dagger}(x_2) \mid 0 > \| _{x_{10} = x_{20}} \tag{89}$$

plays the role of the potential. Using translation invariance, $V^k(x_1, x_2)$ is a function of $x_1 - x_2$ only, and since $x_{10} = x_{20}$, $V^k(x_1, x_2)$ is time independent, and a function of $(\overline{x}_1 - \overline{x}_2)$ only. With our assumption on the B's, Ruelle has shown that $V^k(x_1, x_2)$ falls off faster than any inverse power of $|\overline{x}_1 - \overline{x}_2|$ but not exponentially, for fixed Λ_i. As Λ_i also varies the appropriate falloff parameter becomes $|\overline{x_1} - \overline{x_2}|/E$ where E is the larger of the two energies of the Lorentz transformations Λ_1 and Λ_2[23].

Two features of $V^k(x_1, x_2)$ should be noted. First, it depends on E and corresponds to a potential whose range increases linearly with E. This arises from the Lorentz transformation Λ_i in the definition of the $B_i^{\Lambda_i}$, and is necessitated by the rapid falloff of $g_i(\overline{p})$ with energy. Since the wave packet created by $B_{f_i}{}^\dagger$ is $g_i(\overline{p}) \tilde{f}_i(\overline{p})$, the probability of producing high energy states without the Λ_i is very small and so no effective information can be obtained on high energy scattering. With the Lorentz transformation $g_i(\Lambda_i^{-1}p)$ is concentrated near $\bar{k}_i$ and there is an appreciable probability of creating energetic states. But since g_i vanishes for $p^2 > M_i^2$, (see (64)), $g_i(\Lambda_i^{-1}p)$ is confined to the following small region

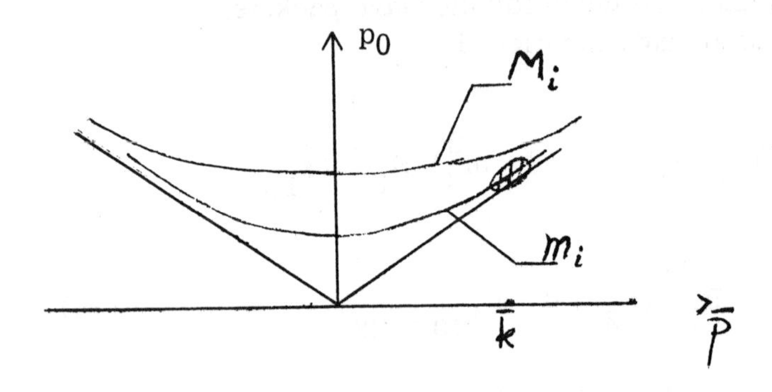

As k increases, the two hyperbolas of mass m_i, M_i approach each other, so that g has to vary rapidly to go from zero at $p^2 = M_i^2$, to something of order 1 for $p^2 = m_i^2$. But rapid variations in p space mean long tails in x space, and so the field $B^\Lambda(x)$ is not well localized, and its effective range increases like E. This dependence on E is therefore a manifestation of relativistic kinematics, for in the non-relativistic case, the mass shells are parabolae, not hyperbolae, and as the energy increases they get further and further apart, so that is is easy to build functions g localized at $p = k_i$, vanishing for $(\bar{p}^2/2p_o) > M$, and without long tails in x space.

Secondly, $V^k(\underline{x}_1, \underline{x}_2)$ does not fall exponentially in $|x_1 - x_2|/E$, but only faster than any inverse power. This arises from the condition that g(p) vanish for $p^2 > M^2$. Thus the Fourier transform of g cannot fall exponentially in x since if it did, its Fourier transform would be analytic in some complex region of p space including real p, and therefore couldn't vanish identically for $p^2 > M^2$. The spectrum condition was imposed on g so that $B^\dagger$ would create only one-particle states. This in turn allowed $V^k(x_1, x_2)$ to fall off rapidly in $|\overline{x}_1 - \overline{x}_2|$. If $B^\dagger$ created many particle states, the various particles created by $B_1^\dagger$ would interact among themselves and the "potential" (89) would not vanish as x_1 and x_2 became separated.

The situation is then very much as in scattering by a potential which falls off faster than any inverse power of the separation, but not exponentially, and which has a range increasing linearly with k. We do not expect to prove any analyticity, nor the Froissart bound, but we can establish the Greenberg-Low bound. To do this we make separate estimates of the left and right sides of (79).

As in the potential case, with our wave packets and bounds on $V^k(x_1, x_2)$, we can show that the right side of (79) falls off faster than any power of a/E. The left side is larger than

$$\int ds \, \mathrm{Im}\, T_l(s)\, F_l(s) \tag{90}$$

for any l, where $F_l(s)$ is the amount of partial wave l in the incident packet $h_1(p_1) h_2(p_2)$ at total center of mass energy = $\sqrt{s}$. Using (80), and choosing $\phi(x_1, \ldots x_n)$ so that $g(\overline{p}) = 1$ near $\overline{p} = 0$ and decreasing away from $\overline{p} = 0$, we can show as in the potential case that $F_l(s)$ is peaked near $s = (k_{10} + k_{20})^2$, and that if $\mathrm{Im}T_l(s)$ is smooth in s, it follows that for $a = l/k$, $\mathrm{Im}T_l(s)$ falls off faster than any inverse power of $l/kE \approx l/k^2$ for large k. We have consistently ignored factors of powers of E and of a. These only add terms of order s^ϵ to the bound, so that finally

$$\sigma(s) < C\, s^{1+\epsilon} \quad . \tag{91}$$

By being more careful, one can actually improve this to

$$\sigma(s) < C\, s(\ln s)^{2+\epsilon} \, . \tag{92}$$

If we had not made the Lorentz transformation, the right side of (79) would have decreased faster than any power of a, but the left side would have been roughly

$$g_1(\bar{k})\; g_2(-\bar{k}) \quad \int ds\; \mathrm{Im}\, T_l(s)\; F_l(s) \tag{93}$$

and since g_i decreases rapidly with k, no useful bound on $\mathrm{Im}T_l(s)$ would have been obtained. This confirms the earlier observation that without the Lorentz transformation, it is difficult to create states of high energy, and so no useful information on high energy scattering can be obtained.

Finally, let me mention that stronger results can be obtained.[24] We failed to prove analyticity because we wanted our fields to create only one-particle states when applied to the vacuum, so that the "potential" would decrease rapidly in $|\bar{x}_1 - \bar{x}_2|$. One can however write the field B_i as

$$B_i = C_i - D_i \tag{94}$$

where C_i creates both single-particle and many-particle states, and D_i subtracts the many-particle contribution. One can then explicitly cancel the interactions among themselves of the many-particle states created by each of the C_i and D_i. The remaining "potential" arising from the particles created only by the C's can be shown to decrease exponentially with $|\overline{x}_1 - \overline{x}_2|$ by using a stronger cluster property derived by Araki, Hepp and Ruelle.[25] The "potential" involving particles created by D_i cannot decrease exponentially, since those arising from B_i do not. However, we can make the contribution of terms involving D_i decrease exponentially with a by making the probability of creating many-particle states decrease exponentially with a. This is accomplished by making $g_i(p)$ a-dependent as well as k-dependent which it already is. One can then prove that the scattering amplitude is analytic in t in an ellipse which is smaller than the Lehmann ellipse but which has the same energy dependence at high energies. Recall that this is for scattering of particles whose interpolating fields are not local.

Finally, let me mention one puzzling feature of this investigation. By using the cluster properties of Ruelle, and of Araki, Hepp, and Ruelle, it seems that we have exploited the short range character of the forces to a great extent. But our result is the Greenberg-Low bound, not the Froissart bound, as we had originally hoped. The question is; What extra physical principles are involved in the Froissart bound? Because of their mathematical complication, it is very difficult to unravel the present proofs of the Froissart bound to lay bare the underlying physics. One possibility is that we have not made good use of Lorentz invariance. You recall that the extra powers of energy in the Greenberg-Low bound arise because mass hyperboloids approach each other for large energy. This is a relativistic kinematic effect, and it is quite possible that our failure to consistently exploit relativistic kinematics has led to the weaker results.

CONCLUSIONS

We have presented a very physical approach to t analyticity and to the high energy bounds. We have also established results for bound state scattering. Some problems still remain, and their resolution may well lead to deeper physical insight about the role of relativistic kinematics in high energy scattering.

APPENDIX

We wish to estimate

$$\int_{-\infty}^{\infty} \| V\varphi(t') \| \, dt' \tag{A1}$$

where

$$|V(r)| < \frac{Ce^{-\mu r}}{r} \tag{A2}$$

and

$$|\varphi(\bar{x}, t)|^2 = \left(\frac{1}{\sqrt{\pi}\, b(t)}\right)^3 e^{-(\bar{x}-\bar{a}-\bar{k}t/m)^2/b(t)^2} \tag{A3}$$

with

$$b(t)^2 = b^2 + t^2/m^2 b^2 \tag{A4}$$

and

$$b^2 = \lambda a \tag{A5}$$

and

$$\bar{k} \cdot \bar{a} = 0 . \tag{A6}$$

Now

$$\begin{aligned} \|V\varphi(t)\|^2 &= \int |V(\bar{x})|^2 |\varphi(\bar{x}, t)|^2 d^3x \\ &= \int_{r<\rho} |V(\bar{x})|^2 |\varphi(\bar{x}, t)|^2 d^3x \\ &+ \int_{r>\rho} |V(\bar{x})|^2 |\varphi(\bar{x}, t)|^2 d^3x \end{aligned} \tag{A7}$$

and we choose

$$\rho < a . \tag{A8}$$

In the region $r > \rho$, we have

$$|\varphi(\bar{x}, t)|^2 < \left(\frac{1}{\sqrt{\pi}\, b(t)}\right)^3 \tag{A9}$$

so that using (A2), the second term in (A7) is bounded by

$$\frac{2C^2}{\sqrt{\pi}\,(b(t))^3} \quad \frac{e^{-2\mu\rho}}{\mu} \quad . \tag{A10}$$

We bound the first term in (A7) by

$$\underset{r<\rho}{\text{Max}}\; |\varphi(\bar{x}, t)|^2 \int_{r<\rho} |V(\bar{x})|^2 d^3x$$

$$\leq \underset{r<\rho}{\text{Max}}\; |\varphi(\bar{x}, t)|^2 \;\frac{4\pi C^2(1-e^{-2\mu\rho})}{2\mu} \quad . \tag{A11}$$

In the region $r < \rho < a$, the minimum value of

$$(\bar{x} - \bar{a} - \bar{k}t/m)^2 \tag{A12}$$

is

$$(\rho - \sqrt{a^2 + k^2t^2/m^2})^2 \tag{A13}$$

where we have used (A6). Inserting this in (A3) gives

$$\|V\varphi(t)\|^2 < \frac{2C^2}{\sqrt{\pi}(b(t))^3\mu}$$

$$\left[e^{-2\mu\rho} + (1 - e^{-2\mu\rho})\, e^{-(\rho - \sqrt{a^2 + k^2t^2/m^2})^2/b(t)^2}\right]. \tag{A14}$$

Using (A5), the maximum of the last term in (A14) as t varies occurs at $t = 0$ providing λ is chosen to satisfy

$$\lambda k \geq 1 . \tag{A15}$$

We then have

$$\|V\varphi(t)\|^2 < \frac{2C^2}{\sqrt{\pi}\,(b(t))^3\mu}\left[e^{-2\mu\rho} + e^{-(\rho - a)^2/\lambda a}\right]. \tag{A16}$$

The bound is best if we choose ρ so that the exponentials are equal, i.e.

$$\rho = a\left[(1 + \lambda\mu) - \sqrt{(1 + \lambda\mu)^2 - 1}\right] \tag{A17}$$

where we have imposed (A8). Using the explicit form (A4), we find

$$\int_{-\infty}^{\infty} dt \, \| V \varphi(t) \| \le \frac{2\sqrt{2}\, m^2 (\lambda a)^{1/2} C}{\pi^{1/4} \mu^{1/2}}$$

$$e^{-\mu a \left[(1 + \lambda\mu) - \sqrt{(1 + \lambda\mu)^2 - 1} \right]} . \qquad (A18)$$

For convenience, we shall take

$$\lambda = 1/k \qquad (A19)$$

which is consistent with (A15). Defining τ by

$$\tau = \mu \left[(1 + \mu/k) - \sqrt{(1 + \mu/k)^2 - 1} \right] \qquad (A20)$$

and using the equation

$$\| T \varphi \| \le \frac{1}{2} \int_{-\infty}^{\infty} \| V \varphi(t) \| \, dt \qquad (A21)$$

we have

$$\| T \varphi \| < \text{Const.} \left(\frac{a}{k}\right)^{1/2} e^{-\tau a} \qquad (A22)$$

so that $\| T \varphi \|$ decreases exponentially with a.

If b^2 were independent of a, then we would have

$$\varphi(\bar{p}, t = 0) = \left(\frac{b^2}{\pi} \right)^{3/4} e^{-b^2 (\bar{p} - \bar{k})^2 / 2} \, e^{-i \bar{p} \cdot \bar{a}} . \qquad (A23)$$

If we defined

$$\varphi_{lm}(p) = \int \varphi(\bar{p})\, Y^{*}_{lm}(\hat{p})\, d\Omega_p \tag{A24}$$

then

$$|\varphi_{00}(p)|^2 = \left(\frac{b^2}{\pi}\right)^{3/2} e^{-b^2(p^2+k^2)}\, 4\pi \frac{\sin^2 p\sqrt{a^2 - k^2b^2}}{p^2(a^2 - k^2b^2)} . \tag{A25}$$

But

$$\|T\varphi\|^2 = \sum_{lm} \int p^2 dp\, |\varphi_{lm}(p)|^2\, \mathrm{Im}\, T_l(p)$$

$$\geq \int p^2 dp\, |\varphi_{00}(p)|^2\, \mathrm{Im}\, T_0(p) \tag{A26}$$

and so for large a, we would have

$$\|T\varphi\|^2 > \left(\frac{b^2}{\pi}\right)^{3/2} e^{-b^2k^2}\, 4\pi \int e^{-b^2p^2} \frac{\sin^2 pa}{a^2}\, \mathrm{Im}\, T_0(p)\, dp \tag{A27}$$

so that if

$$T_0(p) > \epsilon \quad \text{over some finite region} \tag{A28}$$

we would have

$$\|T\,\varphi\,\|^2 > \frac{\text{const.}}{a^2} \text{ for large a} \tag{A29}$$

and so $\|T\,\varphi\,\|$ would not fall exponentially with a.

REFERENCES

1. R. Jost, Helv. Phys. Acta. 20, 256 (1947).
2. N. Hu, Phys. Rev. 74, 131 (1948).
3. W. Schützer and J. Tiomno, Phys. Rev. 83, 249 (1951).
4. N. G. Van Kampen, Phys. Rev. 89, 1072 (1953); Physica 20, 115 (1954) Rev. Mex. di Fis. 2, 222 (1953).
5. R. Kronig, Physica 12, 543 (1946).
6. See e.g. E. P. Wigner, Causality, R-Matrix and Collision Matrix, published in Ref. 7.
7. Dispersion Relations and their Connection with Causality, Scuola Internazionale di Fisica "Enrico Fermi," Varenna, E. P. Wigner, ed. Academic Press, N. Y. 1964.
8. M. Gell Mann, M. L. Goldberger and W. Thirring, Phys. Rev. 95, 1612 (1954).
9. See e.g. M. Froissart, The Proof of Dispersion Relations, published in Ref. 7.
10. R. Omnes, Phys. Rev. 146, 1123 (1966).
11. H. Lehmann, Nuovo Cimento, 10, 579 (1958).
12. O. W. Greenberg and F. E. Low, Phys. Rev. 124, 2047 (1961).
13. M. Froissart, Phys. Rev. 123, 1053 (1961).

14. A. Martin, Nuovo Cimento 42, 930 (1966).
15. M. Kugler and R. Roskies, Phys. Rev. 155, 1685 (1967).
16. E. T. Whittaker and G. N. Watson, A Course of Modern Analysis (Cambridge University Press, New York 1952) 4th ed. p. 323.
17. See for example, A. Martin in Strong Interactions and High Energy Physics, ed. R. G. Moorhouse, Oliver and Boyd, London, 1964.
18. R. Haag, Phys. Rev. 112, 669 (1958).
19. D. Ruelle, Helv. Phys. Acta. 35, 147 (1962).
20. See e.g. R. F. Streater and A. S. Wightman, P. C. T., Spin and Statistics and All That, Benjamin, New York (1964).
21. The A's are symbols of any local field operators. They may involve spinor indices, they may involve Hermatian conjugates, and finally, they need not refer to the same field.
22. F. Riesz and B. Sz-Nagy, Functional Analysis, Ungar, New York (1955), Section 137.
23. For explicit estimates, see S. W. MacDowell, R. Roskies and B. Schroer, Phys. Rev. 166, 1691 (1968).
24. S. W. MacDowell and R. Roskies, Phys. Rev. 166, 1703 (1968).
25. H. Araki, K. Hepp and D. Ruelle, Helv. Phys. Acta 35, 164 (1962).

MESON DYNAMICS

Paul Singer

MESON DYNAMICS

Paul Singer*
Research Institute for Theoretical Physics
University of Helsinki

I. CURRENT ALGEBRA, PCAC AND MESON-MESON INTERACTIONS

1. Introduction

At the previous Liperi Summer School in Theoretical Physics I presented a detailed review on Meson-Meson interactions.[1] At that time, the art of using current algebras and partial conservation of the axial current was "in nascentia" and I refrained from including its contribution. The last two years have been marked by an ever increasing flux of papers using these approaches in various domains of Particle Physics. Some of the more interesting results have to do indeed with meson dynamics. Especially, the pion-pion interaction has been the subject of several analyses and the conclusions of

*Permanent address: Department of Physics, TECHNION-Israel Institute of Technology, HAIFA, ISRAEL.

the various approaches are sometimes at variance. The picture of strong S-wave π-π interaction at low energy, which seemed to hold a strong position a year ago, has been violently questioned lately. The experiments in this area are still unable to provide us with a clear cut answer. The field is therefore widely open to both experimental and theoretical fresh ideas.

2. Consistency Relations

The hypothesis[2] of partial conservation of the axial current (PCAC), has been used already years ago[3] in order to obtain relations connecting the weak and strong interactions. The most notorious of these is the Goldberger-Treiman relation,[4] relating the charged pion life-time, the ratio of the axial to vector weak coupling G_A/G_V and the strong pion-nucleon coupling constant. More recently, Adler[5] has shown that the PCAC assumption can be used to derive consistency conditions involving the strong interactions alone. For instance, he obtains a relation involving the pion-nucleon symmetric isotopic spin amplitude and the renormalized pion nucleon coupling constant g

$$A^{\pi N(+)}(\nu = 0,\ \nu_B = 0,\ k^2 = 0) = \frac{g^2 K_{NN\pi}(k^2 = 0)}{M} \tag{1}$$

where $K_{NN\pi}$ is the pionic form factor of the nucleon and M is the nucleon mass, k^2 is the $(\text{mass})^2$ of the initial pion, ν and ν_B are energy and momentum-transfer invariant variables and the final pion is on the mass shell. By using π-N dispersion relations and pion-nucleon experimental data, Adler finds this relation to be satisfied within 10%. This is approximately also the degree of accuracy of the Goldberger-Treiman relation.

In Adler's approach,[5] the PCAC hypothesis is expressed in a field theoretic sense by the definition

$$\partial_\mu A^\mu = C_\pi \varphi_\pi + R \tag{2}$$

where A_μ is the axial vector part of the strangeness-conserving current and φ_π is the renormalized field operator which creates a π-meson. The interesting relations are obtained for the situations in which for states α, β one has $< | \beta | \varphi_\pi | \alpha > \neq 0$ while

$$< \beta | R | \alpha > \ll C_\pi < \beta | \varphi_\pi | \alpha >$$

and henceforth the contribution of the residual operator R is neglected. The constant C_π is postulated to have the value which would reproduce the Goldberger-Treiman relation, i.e.

$$C_\pi = - \frac{G_A/G_V \, M\mu^2 \sqrt{2}}{g \, K_{NN\pi}(0)} . \tag{3}$$

$K_{NN\pi}(0)$ is the pionic form factor of the nucleon evaluated at zero mass of the pion and normalized to $K_{N\pi\pi}(\mu^2) = 1$, where μ is the pion mass. The rest of the constants appearing in (3) have been defined previously. $\sqrt{2}$ contains the convention about the pion-nucleon coupling constant.

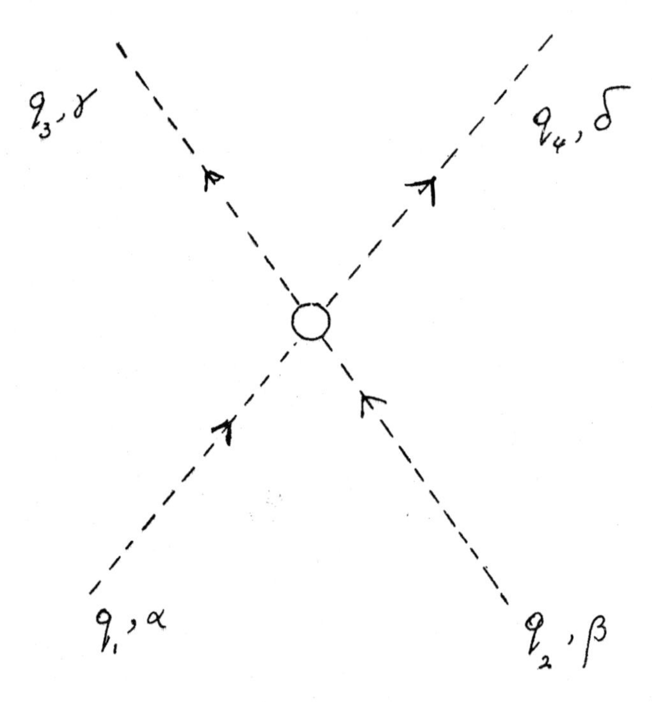

Fig. 1.

Let us discuss here in detail the consistency conditions obtainable for pion-pion scattering. The matrix element for pion-pion scattering is defined with the notation of Fig. 1 where α, β, γ, δ are isotopic spin indices. The isotopic spin structure is

$$\langle \pi\, \pi;\ \text{out} \mid \pi\, \pi;\ \text{in} \rangle = (16\, q_{10}\, q_{20}\, q_{30}\, q_{40})^{-1}$$

$$\psi^*_\gamma\, \psi^*_\delta\, M_{\alpha\beta\gamma\delta}(s, t, u)\, \psi_\alpha\, \psi_\beta \tag{5}$$

with

$$M_{\alpha\beta\gamma\delta} = A(s,t,u)\,\delta_{\alpha\beta}\,\delta_{\gamma\delta} + B(s,t,u)\,\delta_{\alpha\gamma}\,\delta_{\beta\delta} + C(s,t,u)\,\delta_{\alpha\delta}\delta_{\beta\gamma}. \tag{6}$$

The s, t, u invariant variables are defined

$$s = (q_1 + q_2)^2;\ t = (q_1 - q_3)^2;\ u = (q_1 - q_4)^2$$

$$s = (q_3 + q_4)^2;\ t = (q_2 - q_4)^2;\ u = (q_2 - q_3)^2$$

$$s + t + u = 4\,\mu^2. \tag{7}$$

Crossing symmetry and Bose statistics require A to be a symmetric function of t and u, B to be symmetric in u and s and C to be symmetric in t and s and also relate the functions A, B, C when interchanging the three variables.[6]

Consider now the matrix element of the axial current $< m\,|A_\mu|\eta >$ with $< m\,| = \,< q_3, \gamma; q_4, \delta\,;\text{out}\,|$ and $|\eta> = |q_1, \alpha>$ and associate with A_μ the isotopic spin index β and a momentum transfer $k = q_2$.

Defining

$$<q_3\gamma,\ q_4\delta\,|A^\mu|q_1\,\alpha> = (8\,q_{10}\,q_{30}\,q_{40})^{-1}\ \psi^*_\gamma\psi^*_\delta\ M^\mu_{\alpha\beta\,\gamma\,\delta}(s,t,u)\,\psi_\alpha\psi_\beta \tag{8}$$

Adler finds[5]

$$M^{\mu}_{\alpha\beta\gamma\delta}(s,t,u) = \delta_{\alpha\beta}\delta_{\gamma\delta}\Big[- A_1(s;t,u)\,(q_3 + q_4)_\mu - A_2(s;t,u)\,(q_3 - q_4)_\mu + A_3(s;t,u)\,q_{1_\mu}\Big]$$

$$+ \delta_{\alpha\gamma}\delta_{\beta\delta}\Big[A_1(t;u,s)\,(q_1 + q_3)_\mu + A_2(t;u,s)\,(q_1 - q_3)_\mu - A_3(t;u,s)\,q_{4_\mu}\Big]$$

$$+ \delta_{\alpha\delta}\delta_{\beta\gamma}\Big[A_1(u;t,s)\,(q_1 + q_4)_\mu + A_2(u;t,s)\,(q_1 - q_4)_\mu - A_3(u;t,s)\,q_{3_\mu}\Big] \qquad (9)$$

with $A_1(u;t,s)$, $A_3(u;t,s)$ symmetric and $A_2(u;t,s)$ antisymmetric in the variables t, s. Multiplying (9) by q_{2_μ} one obtains

$$q_{2_\mu} M^{\mu}_{\alpha\beta\gamma\delta}(s,t,u) = 0 \qquad (10a)$$

$$\text{when } q_2 \cdot q_1 = q_2 \cdot q_3 = q_2 \cdot q_4 = 0 \qquad (10b)$$

as the A_1, A_2, A_3 amplitudes have no poles. From the definitions (7) one easily sees that (10b) implies

$$s = t = u = k^2 + \mu^2 . \qquad (11)$$

On the other side, since $s + t + u = 3\mu^2 + k^2$, it can be consistent with (11) only when $k^2 = 0$.

Hence, one obtains $< \pi\pi | \partial_\mu A^\mu | \pi > = 0$ when $k^2 = 0$ and $s = t = u = \mu^2$. By using the PCAC hypothesis $\partial_\mu A^\mu = C_\pi \varphi_\pi$ one arrives at the consistency condition

$$M_{\alpha\beta\gamma\delta} (s = m_\pi^2 , t = m_\pi^2 , u = m_\pi^2 ; k^2 = 0) = 0 \tag{12}$$

where k^2 is the mass of one of the pions and the three others are on the mass shell.

Fuchs[7] has also derived a similar consistency condition by using however the PCAC assumption in a dispersion theory sense. One assumes in this approach that $\partial_\mu A^\mu$ is a highly convergent operator, whose matrix elements satisfy unsubtracted dispersion relations in the momentum transfer variable k^2. One then assumes that heavy states give unimportant contributions to the spectral function of $\partial_\mu A^\mu$. Thus for small k^2 and certain values of the other variables of the problem the dispersion relation is dominated by low-lying poles. PCAC is the dominance of the pion pole at $k^2 = \mu^2$, close to $k^2 = 0$.

Let us consider the following expression

$$M_\mu = \int d^4x\, e^{ik \cdot x} < \pi\pi | A_\mu (x) | \pi >. \tag{13}$$

One then takes

$$k^\mu M_\mu = i \int d^4x\, e^{ik \cdot x} <\pi\pi | \partial^\mu A_\mu (x) | \pi > \tag{14}$$

which is obtained from (13) after integrating by parts and setting surface terms equal to zero. Assuming that $k^{\mu}M_{\mu}$ satisfies an unsubtracted dispersion relation in k^2 and that it is dominated by the pole term, Fuchs[7] obtains in the limit of forward scattering the following condition for the pion-pion on the mass shell amplitude (for arbitrary isotopic spin)

$$M^{\pi\pi}(s = 2\mu^2, \quad u = 2\mu^2, \quad t = 0) = 0. \qquad (15)$$

Although this is a relation with all pions on the mass-shell (in Adler's case one pion was off-the-mass-shell), it is still in the nonphysical region. Due to the symmetry of the amplitude in the s, t, u variables, this implies the vanishing of the on-the-mass-shell π-π amplitude at two additional points: ($s = 2\mu^2$, $u = 0$, $t = 2\mu^2$) and ($s = 0$, $u = 2\mu^2$, $t = 2\mu^2$). It is worth remembering that the effective pion-pion coupling constant, presumably non-zero, is the value of the negative π-π scattering amplitude at the symmetry point[6] (which, again, is not in the physical region):

$$-A\left(s = t = u = \frac{4\mu^2}{3}\right) = -B\left(s = t = u = \frac{4\mu^2}{3}\right)$$

$$= -C\left(s = t = u = \frac{4\mu^2}{3}\right) = \lambda. \qquad (16)$$

3. Sum Rules

Fuchs has used[7] condition (15) in a dispersion relation, in order to obtain information on π-π

scattering. He assumes for the π-π forward scattering amplitude $A(\nu)$ a subtracted dispersion relation of the form

$$A(\nu) = A(0) + \frac{\mu\nu}{8\pi^2} \int_{\mu}^{\infty} d\nu' \frac{(\nu'^2 - \mu^2)^{1/2}}{\nu'} \left[\frac{\sigma^{tot}(\nu')}{\nu' - \nu} - \frac{\sigma'^{tot}(\nu')}{\nu' + \nu} \right] \quad (17)$$

where ν is the projectile lab energy and σ^{tot}, σ'^{tot} are the total cross sections in the direct and crossed channels.[8] One should remark that

$$\lim_{\nu \to \infty} (\sigma^{tot}(\nu) - \sigma'^{tot}(\nu)) = 0,$$

by Pomeranchuk theorem. By using $A(0) = 0$ which is just condition (15), one has at threshold ($\nu = \mu$) for the isotopic spin channel I, a_I being the appropriate scattering length:

$$a_I \equiv A_I(\mu) = \frac{\mu^2}{8\pi^2} \int_{\mu}^{\infty} d\nu \frac{(\nu^2 - \mu^2)^{1/2}}{\nu} \left[\frac{\sigma_I(\nu)}{\nu - \mu} - \frac{C_{II'}\sigma_{I'}}{\nu + \mu} \right] \quad (18)$$

where $C_{II'}$ is the crossing matrix. By combining the expressions for $I = 0, 1, 2$ the following relations hold:

$$1/2(a_1 + a_2) = \frac{\mu^2}{8\pi^2} \int_\mu^\infty \frac{d\nu}{\nu} (\nu^2 - \mu^2)^{1/2} \frac{\sigma_1 + \sigma_2}{\nu^2 - \mu^2} > 0$$

$$1/2(a_0 + 2a_2) = \frac{\mu^2}{8\pi^2} \int_\mu^\infty \frac{d\nu}{\nu} (\nu^2 - \mu^2)^{1/2} \frac{\sigma_0 + 2\sigma_2}{\nu^2 - \mu^2} > 0. \quad (19)$$

As $a_1 = 0$ at threshold by Bose statistics, $a_2 > 0$. Fuchs then tries to fulfill the sum-rule (19) by using for σ_0 a contribution given by a scattering length formula and a σ resonance (an S wave π-π resonance at about 400 MeV,[9]), for σ_1 the ρ-meson and for σ_2 again a scattering length formula. It turns out that the sum rule cannot be fulfilled if the contribution to σ_0 comes from a scattering length formula only, even if one allows for a scattering length as high as $a_0 \approx$ two pion Compton wave lengths. On the other side, a σ-meson of mass 400 MeV and width $\Gamma_\sigma = 70$ MeV gives a consistent solution. For a_2 a solution is found for $a_2 \approx 0.05\ \hbar/\mu c$. Hence, this work seems to imply the need for a low energy π-π resonance of the σ-type.

Adler,[10] Kawarabayashi, McGlinn and Wada[11] and Muzinich and Nussinov[12] have also obtained π-π sum rules, by using a different procedure. Namely, they follow the approach used by Adler[13] and Weissberger[14] in calculating the axial-vector coupling constant renormalization in β decay on the basis of certain equal-time commutation relations and PCAC.

Let us define the following axial charges, which are the space integrals of the appropriate time component of the axial vector current

$$ {}^{5}Q(t)^{\pi\pm} = \int d^3x \, (A_0^1(x) \pm iA_0^2(x)) \tag{20} $$

where the upper index represents isotopic spin and we denoted the charge by index $\pi^\pm$ to underline its transformation properties. These charges fulfill the commutation relation

$$ [\,{}^{5}Q^{\pi+}(t),\ {}^{5}Q^{\pi-}(t)\,] = 2I_3 \tag{21} $$

where I_3 is the isotopic spin operator. One takes then (21) between charged one pion states and one uses PCAC in the form

$$ \frac{d\,{}^{5}Q^{\pi^\pm}(t)}{dt} = -\frac{\sqrt{2}\,\mu^2 M\, G_A/G_V}{gK_{NN\pi}(0)} \int d^3x\, \phi_{\pi\pm} \tag{22} $$

which is derivable from the usual PCAC assumption

$$ \partial^\mu A_\mu^i = -\frac{\mu^2 M\, G_A/G_V}{gK_{NN\pi}(0)}\, \phi_\pi^i \ . \tag{23} $$

After several manipulations[15] the following sum rule is obtained

$$ \frac{1}{(G_A/G_V)^2} = \frac{2M^2}{\pi g^2 K_{NN\pi}(0)^2} \int_{2\mu}^{\infty} \frac{\omega\, d\omega}{\omega^2 - \mu^2} \left(\sigma_{\pi^+}^{0,\pi^-} - \sigma_{\pi^+}^{0,\pi^+}\right) \tag{24} $$

where

$$\sigma^{0,\pi^-,+}_{\pi^+}$$

is the total cross section for scattering of a zero mass $\pi^{-,+}$ on a physical π^+ at a center of mass energy ω. It should be observed that the kinematical factor in (24) is large in the low-energy region. The authors of references 10, 11, 12 do all agree that (24) can be fulfilled only by assuming a strong low-energy π-π interaction in S-waves, resonant or not, although they differ slightly in their procedure of estimating σ^0. In Refs. (11) and (12) the off-the-mass-shell correction is largely ignored (i.e. one only uses $\sigma = \sigma^0/K^2_{NN\pi}(0)$), while in Ref. (10) a more refined correction is being made to take into account the fact that one of the pions has zero mass, namely

$$\sigma^{1,I}_0 = K_{NN\pi}(0)^2 \left[\frac{(\omega^2 - \mu^2)^2}{\omega^2(\omega^2 - 4\mu^2)^2} \right]^1 \sigma^{1,I} . \quad (25)$$

In formula (25) l is the orbital angular momentum, I is isotopic spin (=0, 1, 2) and accordingly $\sigma^{l,I}$ is the on-mass-shell partial wave cross section. Apparently, the results obtained by the various authors are not critically dependent on these corrections.

Adler evaluates[10] the right hand side of Eq. (24) by breaking the cross sections into contributions from partial waves and approximating l = I = 1 and l = 2, I = 0 waves by the dominant ρ and f mesons. The explicit formulae used are

$$\sigma^{1,1}(\omega^2) = \frac{3\pi\,\gamma_\rho^2\,(\omega^2 - 4\mu^2)^2}{\omega^2(\omega_\rho^2 - \omega^2) + \gamma_\rho^2(\omega^2 - 4\mu^2)^{3/16}} \tag{26}$$

$$\sigma^{2,0}(\omega^2) = \frac{5\pi\,\gamma_f^2\,(\omega^2 - 4\mu^2)^4}{16(\omega_f^2 - \omega^2)^2\omega^2 + \gamma_f^2(\omega^2 - 4\mu^2)^{5/16}} \tag{27}$$

and the reduced widths γ_f, γ_ρ are related to the experimental ones by

$$\gamma_\rho^2 = \frac{16\,\omega_\rho^4\,\Gamma_\rho^2}{(\omega_\rho^2 - 4\mu^2)^3} \quad ; \quad \gamma_f^2 = \frac{256\,\omega_f^4\,\Gamma_f^2}{(\omega_f^2 - 4\mu^2)^5} \ . \tag{28}$$

It turns that these two contributions amount only to 37% of the total required by the sum rule. By including also an I= 0 S-wave parametrized by a scattering length formula and given by

$$\sigma^{0,0} = \frac{16\,\pi a_0^2}{(\omega^2 - 4\mu^2)\,a_0^2 + \omega^2\,[1 + a_0 H(\omega^2)]} \tag{29}$$

with

$$H(\omega^2) = \frac{2}{\pi\omega}\,(\omega^2 - 4\mu^2)^{1/2}\,\ln\frac{(\omega^2 - 4\mu^2)^{1/2} + \omega}{2\mu}$$

one finds that (25) can be satisfied only if

$$a_0 > 1.3 \left(\frac{\hbar}{\mu c}\right) \quad \text{or} \quad a_0 < -8.5 \left(\frac{\hbar}{\mu c}\right).$$

The authors of ref. (11) find it necessary to consider only the low energy region of the right-hand-side of (25). This is approximated by the contributions of the σ-meson (mass 390 MeV, width 90 MeV) and the ρ-meson and the sum rule is fulfilled within 20%. The contribution of the σ is the dominant one, amounting to about 90%. Muzinich and Nussinov[12] conclude that the sum rule can be fulfilled with a σ-meson plus nonresonating S wave $I = 0$ contribution with scattering length $\simeq 1\ (\hbar/\mu c)$ or without σ and a scattering length for the nonresonating part of $\simeq 2\ (\hbar/\mu c)$.

By generalizing relations like (20) and (21) to include also the strangeness changing current, one can study in a similar fashion reactions with kaons. One deals now [16] with eight vector currents V^i_μ and eight pseudovector currents A^i_μ ($i = 1, \ldots, 8$ is an SU3 index and μ is a Lorentz index). The corresponding charges are defined as

$$Q^i(t) = \int d^3x\, V^i_0(\vec{x}, t) \tag{30}$$

$$^5Q^i(t) = \int d^3x\, A^i_0(\vec{x}, t). \tag{31}$$

These charges are assumed to obey the following commutation relations, which obtain also in a quark model:

$$[Q^i(t),\ Q^j(t)] = if_{ijk}\ Q^k(t) \tag{32a}$$

$$[Q^i(t),\ {}^5Q^j(t)] = if_{ijk}\ {}^5Q^k(t) \tag{32b}$$

$$[{}^5Q^i(t),\ {}^5Q^j(t)] = if_{ijk}\ Q^k(t)\ . \tag{32c}$$

The PCAC assumption needs also generalization, and one makes accordingly the assumption

$$\partial^\mu A_\mu^{K+} = C_K\ \varphi_{K^+} \tag{33}$$

where A_μ^{K+} is the axial current having the quantum numbers of a K^+ meson. The value of C_K can be evaluated by taking (33) between p and Λ or p and Σ^0 states and one obtains[17]

$$C_{K(\Lambda)} = -\frac{G_A^{\Lambda}/G_V(M + M_\Lambda)\ M_K^2}{g_{KP\Lambda}\ K_{KP\Lambda}(0)} \tag{34a}$$

$$C_{K(\Sigma^0)} = -\frac{G_A^{\Sigma^0}/G_V\ (M + M_{\Sigma^0})\ M_K^2}{g_{KP\Sigma^0}\ K_{KP\Sigma^0}(0)} \tag{34b}$$

where g and $K(q^2)$ are the appropriately denoted coupling constants and form factors. In order that (33) be consistent, one must have

$$C_{K(\Lambda)} = C_{K(\Sigma^0)} = C_K\ . \tag{35}$$

From (33), C_K is also obviously related to the rate of $K^+ \to \mu^+ + \nu_\mu$ decay. Mathur and Pandit[18] and Kamal[19] use, in addition to (21), also the commutation relation

$$[{}^5Q^{K^+}(t),\ {}^5Q^{K^-}(t)\ ] = Y + Q \tag{36}$$

taken between π^+ and K^+ states and obtain thus additional sum rules with the aid of (33). By using (21) between K^+ states Mathur and Pandit obtain the sum rule:

$$\frac{1}{(G_A/G_V)^2} = \frac{4M^2}{g^2 K_{NN\pi}(0)^2}\frac{1}{\pi}\int_{\mu+M_K}^{\infty}\frac{\omega d\omega}{\omega^2 - M_K^2}\left[\sigma_{K+}^{0,\pi^-}(\omega) - \sigma_{K+}^{0,\pi^+}(\omega)\right] \tag{37}$$

where

$$\sigma_{K+}^{0,\pi^-}(\omega)$$

denotes the $K^+\pi^-$ total cross section for a zero mass pion at the center-of-mass energy ω. Taking (36) between π^+ states, the same authors obtain another sum rule for π - K scattering:

$$\frac{1}{(G_A^\Lambda/G_V)^2} = \frac{2(M + M_\Lambda)^2}{g_{KP\Lambda}^2 K_{KP\Lambda}(0)^2}\frac{1}{\pi}\int_{\mu+M_K}^{\infty}\frac{\omega d\omega}{\omega^2 - \mu^2}\left[\sigma_{\pi+}^{0,K^-}(\omega) - \sigma_{\pi+}^{0,K^+}(\omega)\right]. \tag{38}$$

The K-π cross section can now be investigated, by requiring self-consistency of the sum rules (37) and (38). The authors use the prescription of Adler (25) to correct for the "zero-mass" particles involved and then assume that the I = 3/2 contributions are negligible. Their conclusion is then that: (a) the K* resonance is far from being sufficient to saturate the sum rules; (b) an inequality is obtained

$$6.8 < g^2_{KP\Lambda} / 8\pi (G_A^{\Lambda}/G)^2 < 13.5$$

if it is assumed that the sum rules are simultaneously saturated by a low energy S-wave K-π resonance. However, the parameters of the required resonance seem somewhat improbable, when considering the existing experimental material; (c) If there is no resonance, it seems that some other fairly strong S-wave interaction is needed in the π-K I = 1/2 channel.

By taking the expectation value of (36) between K^+ states (as well as (33)), a sum rule for K-K scattering is obtained by Kamal[19] and Mathur and Pandit.[18] The sum rule reads:

$$\frac{1}{(G_A^{\Lambda}/G_V)^2} = \frac{(M + M_\Lambda)^2}{g^2_{KP\Lambda} K_{KP\Lambda}(0)^2} \frac{1}{\pi} \int_{2\mu}^{\infty} \frac{\omega d\omega}{\omega^2 - M_K^2}$$

$$\times \left[\sigma^{0,K^-}_{K^+}(\omega) - \sigma^{0,K^+}_{K^+}(\omega)\right]. \qquad (39)$$

Kamal has performed a detailed analysis of this expression. The summation below the K^- threshold is assumed to be dominated by the well established vector mesons ω and ρ, whose couplings are determined by SU(3) symmetry broken by singlet-octet mixing only. Moreover, the φ-meson is assumed to dominate the P-wave I = 0 state above threshold, while the P-wave I = 1 state is neglected. Thus one obtains the expression

$$\frac{1}{(G_A^\Lambda/G_V)^2} = \frac{(M + M_\Lambda)^2}{g^2_{KP\Lambda}} \left[g^2_{\rho K\bar{K}} \left\{ \frac{M_\rho^2 - 2M_K^2}{(M_\rho^2 - M_K^2)^2} \right.\right.$$

$$+ 3 \sin^2\theta \frac{M_\omega^2 - 2M_K^2}{(M_\omega^2 - M_K^2)^2}$$

$$\left. + F(M_\varphi)\frac{3}{2}\cos^2\theta \frac{(M_\varphi^2 - 4M_K^2)^{1/2}}{M_\varphi(M_\varphi^2 - M_K^2)} \right\}$$

$$+ \frac{1}{\pi}\int_{4M_K^2}^{\infty} \frac{\omega d\omega}{\omega^2 - M_K^2} (\sigma_S^{I=0}(\omega)$$

$$\left. + \sigma_S^{I=1}(\omega) - \sigma_{K^+K^+}(\omega)) \right] \qquad (40)$$

where $\sigma_S{}^I$ are the cross sections for K^+K^- S-wave scattering in an isotopic spin state I. $F(M_\varphi)$ is an off-the-mass shell correction factor and all the strong coupling form factors have been neglected, with the expressed hope that their influence is mutually cancelled. Let us now turn to the numerical estimate of (40), not without warning however that the conclusions are marred by the uncertainty concerning the off the-mass-shell continuation. Using

$$G_A{}^\Lambda/G_V = 0.83, \; G_{\Lambda KP}{}^2/4\pi \simeq 14$$

(obtained with an SU(3) mixing parameter f $\simeq$ 0.35), Kamal obtains for (40)

$$1 = 0.22 + \frac{(G_A{}^\Lambda/G_V)^2 \, (M + M_\Lambda)^2}{g_{K\Lambda P}^2} \frac{1}{\pi} \int_{4M_K^2}^{\infty} \frac{\omega d\omega}{\omega^2 - M_K^2}$$

$$\left[\sigma_s^{I=0}(\omega) + \sigma_s^{I=1}(\omega) - \sigma_{K^+K^+}(\omega) \right] . \quad (41)$$

It seems that by parametrizing the σ's by using moderate $K\overline{K}$ interaction will result in very little help to fill the discrepancy between 1 and 0.22. However, if the σ-meson exists, and its coupling to $K\overline{K}$ is taken to be $g_{\sigma K\overline{K}} = g_{\sigma\pi\pi}$ (as for an SU_3 singlet), an additional contribution of $\simeq 1$ is provided for the right hand side of (41). If

$$g_{\sigma K\overline{K}} = \frac{1}{2} g_{\sigma\pi\pi}$$

(as for the isosinglet member of a scalar octet), then the σ contribution reduces to $\simeq 0.25$.

4. Pion Scattering Lengths

By using also the techniques of current commutation relations and the PCAC assumption, Weinberg[20] has derived an expression for the scattering length of a pion off a target with isotopic spin T_t. For a target much heavier than the pion, there seems to be general concensus that the approximations involved in the derivation are justifiable. Moreover, the formula gives very nice results for the π-N scattering, in which case there is experimental material for comparison.[21] Let us review concisely Weinberg's approach: The scattering amplitude on-the-mass-shell is defined in terms of the invariant matrix element M by

$$\langle f, \vec{q}\,\gamma\,|S|\,i, \vec{k}\,\alpha \rangle = \frac{-i\delta^{(4)}(p_i + k - p_f - q)}{(2\pi)^2 (16\, q^0 k^0 p_i^0 p_f^0)^{1/2}}$$

$$|\langle f, q\gamma\,|M|\,i,\ k\,\alpha\rangle|_{q^2 = k^2 = \mu^2} \tag{42}$$

where k_μ, q_μ are initial and final pion four-momenta, α, γ are pion isospin indices, i, f label the initial and final state of the target. Then the S matrix is extended off the mass shell by using the relation between the divergence of the axial-vector current and the pion field:

$$< 0 \,|\partial_\mu A^\mu_\alpha (0)\,|\pi_{q\gamma} > \; \equiv \; C^2_\pi \,\delta_{\alpha\gamma} (2q^0)^{-1/2} (2\pi)^{-3/2} \tag{43}$$

and one has

$$\int d^4x\, d^4y <f\,|T\left\{\partial_\mu A^\mu_\gamma (x),\, \partial_\nu A^\nu_\alpha (y)\right\}\,|i> \; e^{-iq\cdot x}\, e^{-ik\cdot y}$$

$$\equiv \frac{i(2\pi)^4 \delta^{(4)} (p_f + q - p_i - k)\, C^2_\pi}{(q^2 - \mu^2)(k^2 - \mu^2)(2\pi)^3 (4\, p^0_i p^0_f)^{1/2}} <f,\, q\gamma\, |M|\, i,\, k\alpha >. \tag{44}$$

The current commutation relations suggested by the quark-model (as well as the σ-model) are used:

$$\left[A^0_\alpha (y),\, A^\mu_\beta (x)\right] \delta(x_0 - y_0)$$

$$= 2i\,\epsilon_{\alpha\beta\gamma} V^\mu_\gamma (x)\, \delta^{(4)} (x - y)$$

$$\left[A^0_\gamma (x),\, \partial_\mu A^\mu_\alpha (y)\right] \delta(x^0 - y^0)$$

$$= i\,\sigma_{\alpha\gamma}(x)\, \delta^{(4)} (x - y) \tag{45}$$

where we have omitted possible Schwinger terms on

the right-hand side of (45). In the limit q^μ, $k^\mu \to 0$, Weinberg obtains

$$\langle f, q\gamma \,|M|\, i, k\,\alpha \rangle \to M^{(0)}_{f\gamma, i\alpha} - 4\mu^4 \left(\frac{1}{C_\pi}\right)^2 (p \cdot q)\,(T_\pi)_{\gamma\alpha} (T_t)_{fi} + \text{poles} + 0(qq, qk, kk) \qquad (46)$$

and gives arguments which show that the terms $M^{(0)}$ and 0 are approximately of the order $(\mu/M_t)^2$ and can therefore be neglected. Then, near threshold (where also for the S-wave part poles are generally absent) one obtains for the scattering length in the isotopic spin channel I the formula

$$a_I = - \frac{g^2 \mu (G_V/G_A)^2}{8\pi M^2} \left(1 + \frac{\mu}{M_t}\right)^{-1} \left[I(I+1) - I_t(I_t+1) - 2 \right] \qquad (47)$$

after using the Goldberger-Treiman relation for C_π. For π-N scattering, for instance, this gives

$$a_{1/2} = 0.2\,\mu^{-1} \;;\; a_{3/2} = -0.1\mu^{-1}.$$

For π-η scattering, one would have $a_{\pi\eta} = 0$.

Proceeding to π-π scattering, Weinberg expands

the off-mass-shell scattering amplitude to second order in momenta,

$$<q_4\delta,\ q_3\gamma\,|M|\,q_2\beta,\ q_1\alpha> \;=\; \delta_{\alpha\beta}\,\delta_{\gamma\delta}\left[A_0 + B_0(u+t) + C_0 s + \ldots.\right] \qquad (48)$$

$$+\,\delta_{\alpha\gamma}\,\delta_{\beta\gamma}\left[A_0 + B_0(s+u) + C_0 t + \ldots.\right]$$

$$+\,\delta_{\alpha\delta}\,\delta_{\beta\gamma}\left[A_0 + B_0(s+t) + C_0 u + ..\right]$$

where + ... contains terms of fourth and higher order in q_i, i.e. no terms linear in q_i^2. The above expression fulfills the requirements of crossing symmetry and Bose statistics. A_0, B_0, C_0 are constants and s, t, u are defined in Eq. (7). The scattering lengths a_I are (remembering that the physical threshold is at $s = 4\mu^2$, $t = u = 0$)

$$a_0 \simeq -\frac{1}{32\,\mu\pi}\left[5A_0 + 8\mu^2 B_0 + 12\mu^2 C_0\right]$$

$$a_2 \simeq -\frac{1}{32\pi\,\mu}\left[2A_0 + 8\mu^2 B_0\right]. \qquad (49)$$

In the limit $q_1^{\mu} = q_3^{\mu} \to 0$ and $q_2^{\mu} = q_4^{\mu}$ (on the mass-shell), expression (46) is supposed to hold also in this case and reads now

$$<q_4\delta,\ q_3\gamma\,|M|\,q_2\beta,\ q_1\alpha> \;\to\; M^{(0)} - 4\mu^4\left(\frac{1}{C_\pi}\right)^2(q_2\cdot q_3)\times\left[\delta_{\alpha\delta}\,\delta_{\beta\gamma} - \delta_{\alpha\beta}\,\delta_{\gamma\delta}\right]. \qquad (50)$$

The above limits for the momenta, expressed by the invariant variables are $t \to 0$, $s \to \mu^2 + 2q_2 \cdot q_3$, $u \to \mu^2 - 2q_2 \cdot q_3$. Comparing (48) to (50) one has

$$B_0 - C_0 = -\frac{2\mu^2}{C_\pi^2} \tag{51}$$

$$M^{(0)} = \delta_{\alpha\gamma}\delta_{\beta\delta}\left[A_0 + 2\mu^2 B_0\right] + \left[\delta_{\alpha\delta}\delta_{\beta\gamma} + \delta_{\alpha\beta}\delta_{\gamma\delta}\right] \times (A_0 + \mu^2 B_0 + \mu^2 C_0). \tag{52}$$

Using (49) and (51) and the G-T relation one obtains

$$2a_0 - 5a_2 = \frac{3g^2 \mu K^2_{NN\pi}(0)}{4\pi M^2 (G_A/G_V)^2} \simeq 0.69 \frac{\hbar}{\mu c}. \tag{53}$$

As a second step, one assumes that $M^{(0)}$ is proportional to $\delta_{\alpha\gamma}\,\delta_{\beta\delta}$, which could be the consequence of assuming $\partial_\mu A_\alpha{}^\mu$ to be part of a chiral quadruplet, along with an isoscalar field. This gives from (52)

$$A_0 = -\mu^2 (B_0 + C_0). \tag{54}$$

Adler's consistency condition (Eq. 12) is now used with expression (48) for the amplitude and gives

$$A_0 = -\mu^2 (2B_0 + C_0). \tag{55}$$

(55) and (54) taken together imply

$$B_0 = 0 \; ; \; A_0 = - \mu^2 C_0 \tag{56}$$

and hence

$$a_0/a_2 = - \frac{7}{2} \quad . \tag{57}$$

Combining (53) and (57) one obtains

$$a_0 = 0.2 \; \hbar/\mu c \quad ; \quad a_2 = - \; 0.06 \; \hbar/\mu c \tag{58}$$

and the full π-π amplitude reads now

$$< q_4 \, \delta, \, q_3 \gamma \; |M| \; q_2 \beta, \, q_1 \alpha > \; = \frac{2\mu^2}{C_\pi^2} \tag{59}$$

$$\{ \delta_{\alpha\beta} \, \delta_{\gamma\delta} \, (s - \mu^2) + \delta_{\alpha\gamma} \, \delta_{\beta\delta} \, (t - \mu^2) + \delta_{\alpha\delta} \, \delta_{\beta\gamma} \, (u - \mu^2) \} \; .$$

Obviously, (51) points to an a_0 much smaller than one has previously thought.

The results obtained by Weinberg are obviously dependent in a critical manner on the various assumptions made. For instance, if one assumes that $M^{(0)}$ is proportional to

$$(\delta_{\beta\gamma} \, \delta_{\alpha\delta} + \delta_{\gamma\delta} \, \delta_{\alpha\beta})$$

instead of

$\delta_{\alpha\gamma}\,\delta_{\beta\delta}$ as Weinberg did, one obtains[22]

$$a_0 = 0.35\ \hbar/\mu c, \quad a_2 = 0.$$

One should remark however, that such assumption is not consistent with Khuri's consistency condition (Eq. 63), which indeed would give

$$A_0 = -\mu^2 (C_0 + B_0)$$

and

$$A_0 + 2\mu^2 B_0 = -\frac{2\mu^4}{C_\pi^2},$$

exactly as implied by the assumption that $M^{(0)}$ behaves like $\delta_{\alpha\gamma}\,\delta_{\beta\delta}$. Moreover by using PCAC in a dispersion-theory form (see eq. 15 and the discussion preceding it) with the rest of the assumptions as Weinberg's, Fuchs shows that one obtains

$$a_0 = 0.06\,\mu^{-1}, \quad a_2 = -\ 0.12\mu^{-1}.$$

Not the least dangerous approximation in Weinberg's approach is the expansion of the off-mass-shell π-π scattering amplitude in a power series to second order in momenta. In order to make contact with the physical scattering lengths, it is assumed that the expansion is good up to and somewhat beyond threshold, although it is explicitly obvious that it does not have the correct behaviour, required by unitarity, on the physical cut. Khuri[23] has investigated the behaviour of the power series expansion of the π-π amplitude up to second order terms. In his work he uses the commutation relations of the axial-vector charge with the scalar and pseudoscalar densities u_i, v_i as postulated by Gell-Mann:[16]

$$\left[{}^5Q_i(t),\ v_j(\vec{x}, t) \right] = i d_{ijk}\, u_k\, (\vec{x}, t)$$

$$\left[{}^5Q_i(t),\ u_j(\vec{x}, t) \right] = -\, i d_{ijk}\, v_k(\vec{x}, t) \qquad (60)$$

$$i, j, k = 0, 1, \ldots, 8 \ .$$

A generalized Adler consistency condition is obtained from the off-the-mass-shell invariant $\pi\pi$ amplitude

$$M(q_4\delta,\ q_3\gamma,\ q_2\beta,\ q_1\alpha \left[C_\pi a_\pi (2\pi)^3 (4q_4^0\, q_2^0)^{1/2} \right]$$

$$= (q_3^2 - \mu^2)(q_1^2 - \mu^2) \int d^4x\, e^{-iq_1x}$$

$$< \pi_\delta\, (q_4)\, | T(\partial_\mu A^\mu_\alpha(x)\ v_\gamma(0))\, | \pi_\beta(q_2) > \qquad (61)$$

where a_π is a normalization constant defined

$$< 0\, | v_\alpha(0)\, | \pi_\beta\, (q) > = (2\pi)^{3/2}\, (2q^0)^{1/2}\, \delta_{\alpha\beta}\, a_\pi, \qquad (62)$$

which drops from the final result.

Integrating (61) by parts, using the commutation relations (60) and letting q_1, $q_3 \to 0$, Khuri obtains for

A, B, C (Defined in eq. 6) the following consistency condition:

$$A\,(s = \mu^2,\ t = 0,\ u = \mu^2,\ q_1^2 = 0,\ q_2^2 = \mu^2,\ q_3^2 = 0,\ q_4^2 = \mu^2) = 0$$

$$B\,(\mu^2,\ 0,\ \mu^2;\ 0,\ \mu^2,\ 0,\ \mu^2) = -\ 2\mu^4/C_\pi^2$$

$$C\,(\mu^2,\ 0,\ \mu^2;\ 0,\ \mu^2,\ 0,\ \mu^2) = 0 \quad . \tag{63}$$

Khuri also derives consistency conditions for the off-the-mass-shell scattering amplitude which hold not only at one point, but in a three-dimensional s, t, u region. These conditions read:

$$A\,(s, t, u;\ q_1^2 = 0,\ q_2^2 = s,\ q_3^2 = t,\ q_4^2 = u) \simeq 2\mu^2\,(s - \mu^2)/C_\pi^2$$

$$B\,(s, t, u;\ q_1^2 = 0,\ q_2^2 = s,\ q_3^2 = t,\ q_4^2 = u) \simeq 2\mu^2\,(t - \mu^2)/C_\pi^2$$

$$C\,(s, t, u;\ q_1^2 = 0,\ q_2^2 = s,\ q_3^2 = t,\ q_4^2 = u) \simeq 2\mu^2\,(u - \mu^2)/C_\pi^2 \tag{64}$$

and hold in the three dimensional region $0 \leq s$, t, $u \leq \mu^2$. The $\simeq$ sign has to do with the approximation involved in the σ term, by assuming no low-lying S-wave π-π resonance.

Then the A, B, C amplitudes are expanded as follows

$$A(s, t, u; q_1^2, q_2^2, q_3^2, q_4^2) = a + b(t + u) + cs + d(t + u)^2 + etu + fs^2 + g(t + u)s + h\sum_{i>j} q_i^2 q_j^2 \tag{65}$$

and the consistency conditions (64) are used to estimate the involved coefficients.[24] The result is

$$a = 2\mu^4/C_\pi^2 ;\ b = d = f = 0;\ \ c = -2\mu^2/C_\pi^2;$$
$$h = -e = -g \tag{66}$$

The coefficient h is therefore undetermined by the consistency conditions. However arguments are advanced to show that h is small, and taking $h \simeq 0$ and assuming the expansion (65) to hold slightly above $s > 4\mu^2$, one obtains new values for the scattering lengths. These turn out to be equal to Weinberg's (Eq. 58) within 4%.

Sucher and Woo[25] have investigated the consequences of the assumption made by both Weinberg and Khuri, that the power series expansions (48) and (65) are still valid beyond threshold. They conclude, that although the solution of Weinberg and Khuri (W.-K.) is self-consistent, other self-consistent solutions are also possible, and they differ significantly from the W.-K. solution. For instance, they assume an amplitude

$$A = a + b(t + u) + cs + x(s - 4\mu^2)^{1/2} \qquad (67)$$

and they use Khuri's consistency conditions as well as unitarity for the $l = 0$ partial wave in the $I = 0$ channel. Then two solutions are possible,

(a) $a_0 = 0.2\ (\hbar/\mu c)$;

(b) $a_0 = 1.95\ (\hbar/\mu c)$.

Hence, a slight modification of the power-series allowing for unitarity at threshold brings in a solution with fairly large a_0.

II. RADIATIVE DECAYS

1. Current Algebra Approach

The current commutation relations generating the algebra $U(2) \otimes U(2)$ in combination with the PCAC assumption have yielded some good results also in the domain of electromagnetic decays. Although most of these results have been previously obtained with the use of a specific model (namely, the vector dominance model), the successes of the new approach have produced some complimentary understanding. In addition the use of the current algebra + PCAC in radiative decays has much relevance in investigation of the limits of validity of the soft pion approximation.

Kawarabayashi and Suzuki[26] have done the pioneering work by relating the amplitude for A → B + "pair of P-wave pions" to the amplitude for A → B + "isovector photon". The amplitude for the process $A \to B + \pi^i + \pi^j$ is expressed as (i, j, are isotopic spin indices):

$$T = \langle B(k)\pi^i(q_i)\pi^j(q_j);\ \text{out}/A(p);\ \text{in}\rangle$$

$$= -(2\pi)^{-3}(4q_{i0}\,q_{j0})^{-1/2}\int d^4x\,d^4y\ \exp[-i(q_i x + q_j y)]$$

$$K_x K_y \langle B;\ \text{out}\,|T[\varphi^i(x),\ \varphi^j(y)]\,|A;\ \text{in}\rangle \qquad (68)$$

where $\varphi^i(x)$ is the renormalized pion field and K_x is the Klein-Gordon operator

$$(= -\Box_{x^2} + \mu^2).$$

Using now PCAC (Eq. 23) as well as integration by parts, they obtain

$$T = \frac{2}{(2\pi)^3 (4q_{i0}\, q_{j0})^{-1/2} C_\pi^2}$$

$$\{ \int d^4x \int d^4y \exp(-iq_i x - iq_j y)\, K_x K_y\, [\delta(x_0 - y_0)$$

$$< B;\ \text{out} \,|\, [A_0^i(x), \partial^\mu A_\mu^j(y)] \,|\, A;\ \text{in} >$$

$$+ iq_\nu^i\, \delta(x_0 - y_0) < B;\ \text{out} \,|\, A_\nu^i(x),\ A_0^j(y) \,|\, A;\ \text{in} >$$

$$- q_\nu^i q_\mu^j < B;\ \text{out} \,|\, T(A_\nu^i(x), A_\mu^j(y)) \,|\, A;\ \text{in} >]\} . \tag{69}$$

In the limit q^i, $q^j \to 0$, and keeping terms up to first order in each pion momentum one has

$$\lim_{q^i, q^j \to 0} (2\pi)^3 (4q_{i0}\, q_{j0})^{1/2}\, T = 4\pi\delta(E_A - E_B)\, C_\pi^{-2} \mu^4 \times$$

$$\{< B;\ \text{out} \,|\, [{}^5Q_0^i(0),\ {}^5Q_0^j(0)] \,|\, A;\ \text{in} > \tag{70}$$

$$+ iq_\mu^i < B;\ \text{out} \,|\, [{}^5Q_\mu^i(0),\ {}^5Q_0^j(0)] \,|\, A,\ \text{in} > \}$$

where the generalized axial charges ${}^5Q^i{}_\mu(t)$ are defined

$${}^5Q_\mu^i(t) = \int d^3x\, A_\mu^i(x) . \tag{71}$$

The charges $^5Q^i(t)$ previously used, e.g. in Eq. (60), should read $^5Q_0{}^i(t)$ in this notation. Noticing that the first term in (70) is symmetric while the second one antisymmetric in isospin indices, two relations are obtained from (70) for the amplitudes A → B + "S wave pions" and A → B + "P wave pions" respectively:

$$\lim_{q^i, q^j \to 0} (2\pi)^3 (4q_{i0}\, q_{j0})^{1/2}\, T_{A \to B + \{\pi^i, \pi^j\}} \tag{72}$$

$$= i\, 4\pi\, \delta(E_A - E_B) C_\pi^{-2}\, \mu^4 < B;\text{ out } | u_0(0) | A;\text{ in } >$$

and

$$\lim_{q^i, q^j \to 0} (2\pi)^3 (4q_{i0}\, q_{j0})^{1/2}\, T_{A \to B + [\pi^i, \pi^j]}$$

$$= -(2\pi)\, \delta(E_A - E_B) C_\pi^{-2}\, \mu^4 (q^i - q^j)_\nu$$

$$< B;\text{ out } | Q_{k\nu}(0) | A;\text{ in } >. \tag{73}$$

In order to obtain (72) and (73) use is being made of the following commutation relations:

$$[\,^5Q^i_\mu(0),\ ^5Q^j_0(0)\,] = i\epsilon_{ijk}\, Q^k{}_\mu(0) \tag{74}$$

$$[\,^5Q^i_0(0),\ ^5\dot{Q}^j_0(0)\,] = i\delta_{ij}\, u_0(0)\,. \tag{75}$$

Similarly, the generalized charges of the vector current are defined as

$$Q^j_\mu (t) = \int d^3x\, V^j_\mu (x) \ . \tag{76}$$

Equation (73) relates now the transition $A \to B +$ "P wave pions" to the transition between $A \to B$ caused by the isovector current. Kawarabayashi and Suzuki have used[26] this equation to relate the $\omega \to \pi\gamma$ to the $\omega \to 3\pi$ amplitude. The ratio of the decays is obviously determined by using the experimental value of C_π. The result obtained is

$$\frac{\Gamma\,(\omega \to \pi\gamma)}{\Gamma\,(\omega \to \pi^+\pi^-\pi^0)} \simeq 17\% \tag{77}$$

in fair agreement with the experimental value of 11%[27] as well as with the previous theoretical estimate of Gell-Mann, Sharp and Wagner[28] who used a vector dominance model to evaluate it.

For the example just described, the A and B states were respectively the ω and π. If one takes for A an eta meson and for B a γ, a relation is obtained between the $\eta \to \pi\pi\gamma$ and $\eta \to \gamma\gamma$ (isovector part) amplitudes. Ademollo and Gatto[29] have performed this calculation and obtained

$$\frac{\Gamma\,(\eta \to \pi^+\pi^-\gamma)}{\Gamma\,(\eta \to 2\gamma)} = 0.21 \tag{78}$$

This is to be compared with the older "ρ-dominance" calculation of Brown and Singer[9] which gave 0.13 for the above ratio and the experimental result of 0.15.

The problem of radiative decays has also been attacked by Pasupathy and Marshak,[30] who use a somewhat different approach. They start by expanding the time-ordered product

$$T\{\partial^{\mu} A^{i}_{\mu}(x),\ \partial^{\nu} A^{j}_{\nu}(y),\ V^{k}_{\tau}(0)\} \tag{79}$$

where V_{τ}^{k} is the isovector current. There are seven terms in this expansion, and most of them are dropped when taking it between vacuum and pseudoscalar meson states by using either plausible argumentation or rigorous conservation laws (e.g. parity conservation in e.m. interactions), and Lorentz covariance. Finally, a relation similar to (73) is obtained.

The same approach has been followed by Conway[31] to discuss the $\omega \rightarrow 2\pi + \gamma$ decay. In this decay, the two pions are required by C-conservation to be in an even angular momentum state. After some manipulations, the amplitude is obtained to be proportional to

$$< 0 | V_{\mu}^{3}(0) \ | \omega >,$$

which vanishes because it represents an isovector transition between an isoscalar meson and the vacuum. Thus, the decay $\omega \rightarrow \pi\pi\gamma$ turns out to vanish in the soft pion limit. Of course, as the energies of the pions are fairly sizable in this case, one may wonder what the soft pion approximation exactly means. However, if the non-vanishing part is indeed due to the neglected terms proportional to pion momenta, it is indeed somewhat surprising that terms of the same order do not seriously affect the $\omega \to \pi\gamma/\omega \to 3\pi$ ratio. It should be added that a vector dominance calculation[32] of this decay, by assuming it proceeds through $\omega \to (\rho) + \pi \to \pi\gamma + \pi$, gives a nonvanishing rate related to the $\rho\pi\gamma$ vertex. By using for this vertex the SU (6) value ($< \rho \mid \pi\gamma > \; = 3^{-1} < \omega \mid \pi\gamma >$) one expects[32] $(\omega \to \pi^+ \pi^- \gamma/\omega \to \pi^+\pi^-\pi^0) \simeq 1.5 \times 10^{-4}$. At present, only an upper limit of 5×10^{-2} has been established for this ratio experimentally.

2. Connection Between Current Algebra and Vector Meson Dominance

It becomes obvious from the above discussion that the current algebra describes successfully those processes in which an antisymmetric pion pair is emitted, while the treatment of the emission of an even-wave two pion is plagued with some difficulties. In this connection, it is instructive to reveal the relation between the current algebra and the vector-meson dominance approaches.

Let us define the coupling constants of ρ to two pions ($g_{\rho\pi\pi}$) and to an isovector photon (f_ρ^{-1}) by the following expressions:

$$< \pi^i \pi^j \mid \rho > = -(2\pi)^4 \, \delta^{(4)} (p^\rho - q^i - q^j)(8p_0^\rho q_0^i q_0^j)^{-1/2} \times \epsilon_{ijk} \, \rho_\mu^k \, (q^i - q^j)^\mu \, g_{\rho\pi\pi} \tag{80}$$

$$< 0 \mid V_\mu^3 (0) \mid \rho > = (2p_0^\rho)^{-1/2} \, \rho_\mu^3 \, m_\rho^2 / f_\rho \tag{81}$$

where ρ_μ is the polarization vector of the ρ-meson. Using[26] Eq. (73) with A, B being respectively the ρ meson and the vacuum, one obtains the relation

$$g_{\rho\pi\pi} = \frac{\mu^4 m_\rho^2}{C_\pi^2 f_\rho} . \tag{82}$$

With the additional assumption of the universal coupling of ρ to the isospin[33] (in fact, ρ dominance of the pion form factor is sufficient), one obtains

$$\frac{g_{\rho\pi\pi}^2}{4\pi} = \frac{m_\rho^2 \mu^4}{4\pi\, C_\pi^2} . \tag{83}$$

From the known width of the ρ-meson decay and the π^+ decay rate the evaluation of (83) gives[34] 2.4 - 2.7. for the left-hand side and 2.66 for the right-hand side. Hence, the ratio of $g_{\rho\pi\pi}/m_\rho$ is very well accounted for, by the current algebra and PCAC. Now, as we have seen in the previous section, the current algebra value for the ratio of the amplitudes $A \to B + \gamma_V$ to $A \to B + [\pi, \pi]$ is

$$\frac{M\,(A \to B + \gamma_V)}{M\,(A \to B + [\pi, \pi])} = \frac{eC_\pi^2}{\mu^4} . \tag{84}$$

The ρ-dominance model[28] for the same ratio gives

$$\frac{M(A \to B + \gamma_V)}{M(A \to B + [\pi, \pi])} \simeq \frac{e\, m_\rho^2}{g_{\rho\pi\pi}^2} \qquad (85)$$

provided the invariant mass of the two pions is much smaller than the ρ-mass. Eq. (83) explains henceforth why the current algebra and the vector meson dominance give similar results for certain decay ratios. This point has been emphasized by Ademollo[35] and Sakurai.[34]

Some remarks should be made at this stage. The current algebra calculations refer indeed only to soft pions and the question arises as to how far the extrapolation to physical ones can be made. For instance, in the decay $\eta \to \pi\pi\gamma$ or better $X^0 \to \pi\pi\gamma$, one can look at the invariant mass spectrum of the pions to detect whether a general matrix element for the amplitude is sufficient,[29,30] or alternatively, the effect of the virtual ρ-meson is detectable.[9,34] Although the experimental statistics is still low, an analysis of the $\pi^+\pi^-$ spectrum in 33 events of $\eta \to \pi\pi\gamma$ by Crawford and Price[36] gives better agreement with the ρ-dominance model. For the $X^0 \to \pi^+\pi^-\gamma$ decay, where the $\pi^+\pi^-$ can be in the region of the physical ρ-mass, the soft pion approximation is expected to be poor. Indeed the ρ-mass shows up clearly in the spectrum,[37] and a current algebra calculation for the rate appears to be unsuitable.[38]

The current algebra results agree with the ρ-dominance model also in the calculation of scattering lengths.[34] For instance, formula (47) is shown to be equivalent to the result obtained by calculating s-wave π-N scattering through ρ-exchange. However,

(47) is obviously more general, as it applies also to situations where no ρ-exchange is permitted.

III. η-MESON PHYSICS

1. Introduction

Although six years have passed since the discovery[39] of this isoscalar meson, the complexity of its decays and the many puzzles they offer have kept its study constantly in the forefront of particle physics.

Among the many mesonic states discovered since 1960, η is unique in its incapability of decaying strongly. Its known decay modes are apparently of electromagnetic character. However, due to its heavy mass (548.6 ± 0.4 MeV/c^2), this heavier cousin of the π^0 has many possible decays - and the understanding of their relative rates is still a largely unsettled question.

The mere intrinsic properties of η (isotopic spin, mass, spin and parity) make it a very interesting object to study in connection to various open physical problems. For instance, its decays might be a good source for studying possible C-violation in electromagnetic interactions,[40,45,52] while its production by neutrinos could help in investigating[41] CP violation in weak interactions.

In these lectures we shall mainly discuss the various decay modes of the η-meson. In addition we shall touch briefly on the question of the η life-time.

2. Decay Scheme

The η-meson is observed to decay into two γ-rays, which establishes that it is a meson of positive C, with spin 1 excluded. Hence the spin is probably even (if $J \geq 3$ is neglected). Being also an isospin singlet, its G parity ($G = C \exp(i\pi T_2)$) is positive. As no two pion decay is observed, the η-meson parity must be negative. The possible quantum number assignments (with the notation J^{PG}) are then 0^{-+}, 2^{-+}, ... It is generally believed, and also consistent with the existing experimental evidence, that the η is a pseudoscalar meson, belonging to an SU_3 octet together with the pion triplet and the kaon doublet. Its mass comes within a few percent close to the value predicted by the Okubo-Gell-Mann square mass formula. Therefore, the question whether it is a pure octet member, or a mixture of a singlet and octet (the X_0 being its counterpart), is largely an open question.

With the 0^{-+} quantum numbers, the lowest final state to which η could decay strongly is a four-pion-state. As the parity and Bose statistics requirements prevent this of being a $4\pi^0$ state and energy conservation forbids η decay to any other four-pion state, the strong decay of η is strictly forbidden.

The electromagnetic decays of η occur as radiative as well as nonradiative processes. In Table I we list the various η decays under three categories: radiative, nonradiative and leptonic (i.e. those involving a lepton pair with or without photons). After each decay the order in

$$\alpha\left(= \frac{e^2}{4\pi\hbar c}\right)$$

TABLE I:

Radiative			Leptonic		
$\pi^+\pi^-\gamma$	; α	; $(4.6 \pm 0.8)\%$	$\pi^+\pi^-e^+e^-$	; α^2	; $(0.1 \pm 0.1)\%$
$\pi^+\pi^-\pi^0\gamma$	; α	; $< 0.2\%$	$\pi^+\pi^-\mu^+\mu^-$	; α^2	; –
$\gamma\gamma$	; α^2	; $(31.4 \pm 2.2)\%$	$\pi^+\pi^-\pi^0e^+e^-$	; α^2	; –
$\pi^0\gamma\gamma$	; α^2	; $(20.5 \pm 3.5)\%$	γe^+e^-	; α^3	; –
$\pi^+\pi^-\gamma\gamma$	; α^2	; $< 0.2\%$	$\gamma\mu^+\mu^-$	; α^3	; –
Non-Radiative			$\pi^0e^+e^-$	; α^4	; –
$3\pi^0$	; α^2	; $(21.0 \pm 3.2)\%$	$\pi^0\mu^+\mu^-$	; α^4	; –
$\pi^+\pi^-\pi^0$	; α^2	; $(22.4 \pm 1.8)\%$	$\mu^+\mu^-$	; α^4	; –
			e^+e^-	; α^4	; –
			$e^+e^-e^+e^-$	; α^4	; –
			$e^+e^-\mu^+\mu^-$	; α^4	; –
			$\mu^+\mu^-\mu^+\mu^-$	; α^4	; –

of the partial decay width is specified, as well as the experimental branching ratio, whenever information is available. The order in α is given within the usual picture of C-conserving electromagnetic interactions. The experimental data are taken generally from Ref. 27, except for the upper limits for the modes $\pi^+\pi^-\pi^0\gamma$ and $\pi^+\pi^-\gamma\gamma$, which were obtained recently by Price and Crawford.[42] There has been another recent determination[43] of the branching ratios among the neutral η decays which was not included in computing Table I and which gives

$$\gamma\gamma : \pi^0\gamma\gamma : 3\pi^0 \simeq 58\% : 24\% : 18\%$$

compared to 43%: 28%: 29% resulting from Table I. As the uncertainties in these numbers run around $\simeq$ 15-20%, no drastic changes will occur in Table I when including these new results.

Glancing at Table I one sees immediately some of the puzzles with which one is faced in η decay, and some others will show up as we go along. The three pion decay modes occur more frequently than the $\pi^+\pi^-\gamma$ mode, although they are one order in α higher and there is the same number of particles in the final state in both cases. Or if to put it differently, the 3π and 2γ occur with the same frequency, although they would be expected to behave like the ratio of three body/two body phase space, which is again of the order of 1/100.

The $\pi^0\gamma\gamma$ decay mode is more frequent than $\pi^+\pi^-\gamma$, although again one order in α higher and again the same number of particles in the final state is involved for both decays.

The ratio of $\eta \to 3\pi^0/\eta \to \pi^+\pi^-\pi^0$, which is experimentally of the order of unity, represents another of the strange features of η-decay.

In order to explain in a rough manner, the occurrence of the various decay modes, Bronzan and Low[44] have introduced the A-quantum number. The various mesons are eigenstates of the A-operator, with eigenvalues ±1. A-violation in a strong or electromagnetic decay is supposed to inhibit it by a factor $\epsilon \simeq 0.01$. The pseudoscalar octet is given the value A = -1 and γ is given the value A = +1. By using now the fact that two body to three body phase space involves another factor of a $\simeq$ 100, one could roughly account for the fact that the A-allowed processes $\eta \to 3\pi$, $\eta \to \pi^0\gamma\gamma$ are as frequent as the A-forbidden ones $\eta \to \pi\pi\gamma$, $\eta \to \gamma\gamma$. On the other hand, if one discusses $\eta \to \pi^+\pi^-\pi^0\gamma$ along the same lines, using again a factor a for the three- to four-particle phase space, one would have[45]

$$(\eta \to \pi^+\pi^-\gamma) \;/\; (\eta \to \pi^+\pi^-\pi^0\gamma) \simeq \epsilon a \simeq 1 \tag{86a}$$

or

$$(\eta \to \pi^+\pi^-\pi^0\gamma) \;/\; (\eta \to \pi^+\pi^-\pi^0) \simeq \alpha a \simeq 1 \tag{86b}$$

Experimentally,[42] the ratio (86b) turns out to be less than 0.009. The A-quantum number assignments run into trouble also when discussing other meson decays or phenomena, and moreover one encounters difficulties when trying to give a theoretical foundation to this scheme.

3. Non-radiative Decays

The three-pion modes are probably the most puzzling ones among eta-decays. By a rough comparison, we indicated that they were expected to occur less frequently than the $\pi^+\pi^-\gamma$ or 2γ decay modes. The observed relative frequencies could be caused by an inhibition of the radiative decays (as suggested by the A-quantum number considerations) or by an enhancement of the 3π-decay mode. Let us make the following simple argument, to get some physical feeling on this question: as the simplest final state is with the three pions in S-waves, one could use in first approximation to estimate the decay rate a structureless matrix element with strength αG as in Fig. 2. The matrix element for decay being of order α, one expects G to be of order 1, unless some

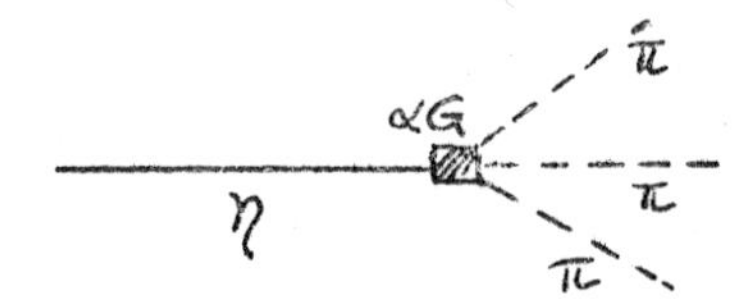

Fig. 2.

unusual effects are at work. The decay rate is then given by

$$\Gamma_{\eta \to \pi^+\pi^-\pi^0} = \frac{\alpha^2 G^2}{(4\pi)^3 m_\eta} \int_\mu^{\omega_{max}} d\omega$$

$$\times \frac{(m_\eta^2 - 2\omega m_\eta - 3\mu^2)^{1/2} (\omega^2 - \mu^2)^{1/2}}{(m_\eta^2 - 2\omega m_\eta + \mu^2)^{1/2}}$$

$$\omega_{max} = \frac{m_\eta^2 - 3\mu^2}{2m_\eta}, \qquad (87)$$

which gives numerically $\Gamma_{\eta \to \pi^+\pi^-\pi^0} = 2.2G^2$ eV. The experimental value for this partial width is apparently of the order of 0.5-1 KeV, involving a value $G^2 \sim 200\text{-}500$. It appears therefore that this decay is enhanced from its "natural" value.

The isotopic spin structure of the three pion states is as following.[46] One can build seven isotopic spin wave functions for three-pion states: one with I = 0; three with I = 1 (one symmetric and two with no determined symmetry); two with I = 2; one with I = 3. The states I = 0, 2 could be attained in η decay only by a C-violating transition (this is easily seen by using the formula $G = C\,(-1)^I$ valid for states characterized by $|I; I_3 = 0>$). Hence, one usually assumes that the $\eta \to 3\pi$ proceeds as a second order electromagnetic transition with

$$|\Delta \vec{I}\,| = 1.$$

The T = 3 final state can be obtained only by a fourth order electromagnetic transition, unless a yet unknown

$$|\Delta \vec{T} = 3\,|$$

interaction with strength of the order 0.01 is at work in this decay.[47]

If one makes the assumption that the three pion states in η decay belong to the symmetric isotopic spin I = 1 wave function, one expects a ratio:

$$R = \frac{\eta \to 3\pi^0}{\eta \to \pi^+\pi^-\pi^0} = \frac{3}{2} \times 1.13 = 1.7 \qquad (88)$$

where 3/2 is a Clebsch-Gordan coefficient and 1.13 is a phase space factor allowing for π^+-π^0 mass difference. The experimental value[27] of $R_{exp} = 0.94 \pm 0.20$ is in strong disagreement with this value.

The Dalitz plot of η decay is very similar to the one encountered in K_2^0 decay. Namely, if one plots the energy distribution of the neutral pion, a preponderance of low energy pions is encountered.

The deviation in Dalitz plot from the expected behaviour for a constant matrix element as well as the low R-value could come about in two ways. It could be due to a strong final state interaction in the S-wave two pion system, or it might be inherent to the $\eta \to 3\pi$ decay mechanism. The first possibility was suggested five years ago by Brown and Singer[48] and shown to be consistent[49] in general lines also with the Dalitz plots of K-meson decay. By assuming an S-wave π-π resonance, the σ-meson, the R as well as the energy spectrum of π^0 are obtained as functions of the mass and width of σ. The best parameters to fit the existing spectrum data, namely $m_\sigma \simeq 400$ MeV, $\Gamma_\sigma \approx 95$ MeV give a ratio $R = 1.19$, about 20% larger than the experimental average.

The π^0-spectrum in the σ-model is a quadratic function of the pion energy E_{π^0}; however, due to the parameters of σ, the quadratic effect is significant only at very low E_{π^0}, the spectrum being otherwise very close to a straight line. Hence, a linear approximation can fit the spectrum as well, and a very large sample of η's is needed in order to differentiate between these two possibilities.

The main drawback of the σ-model is the uncertainty as to whether such a π-π interaction indeed exists. The experimental situation is still inconclusive at present. If the σ turns to be a true feature, then this model has the quality of explaining at the same time R, the spectrum and the large observed partial width for $\eta \to 3\pi$. This last point is obviously due to the rate enhancement caused by the resonating pion pair. In this respect, one should remark that a σ with very large width (as deduced for instance, by Donnachie et al[50] from analysis of backward π-N dispersion relations) will be of no use in accomplishing the above mentioned agreement for η-decay.

It should also be added, that if a scattering length approximation is used for the π-π interaction, one finds it difficult[51,52] to bring R below 1.4-1.5, even if using scattering lengths as high as 1.5 $(\hbar/\mu c)$.

Lately, the current algebra has been applied also to this problem. The sign of alarm was sounded by D. Sutherland[53] who showed that the $\eta \to 3\pi$ matrix element vanishes at certain unphysical points and if a linear matrix element extrapolation is permissible, this implies the forbiddness of this decay mode.

In fact, such an alarming result has been obtained with a different model years ago. Due to the similarities of the η, K $\to 3\pi$ decays, one of the favored dynamical models suggested for these decays is the pion pole model.[51,54,55] The decays would proceed through the sequence η, K $\to \pi \to 3\pi$. Alas, when using the SU(3) invariant four-boson interaction

$$L_{int} = 4\pi\lambda\,(\eta\eta + \vec{\pi}\cdot\vec{\pi} + 2\overline{K}\cdot K) \tag{89}$$

one has two diagrams contributing to $\eta \to 3\pi$ (see Fig. 3) which add to give a null result for the $\eta \to 3\pi$ amplitude.[56] This, of course, neglecting the momentum dependence of the vertices involved. If one allows

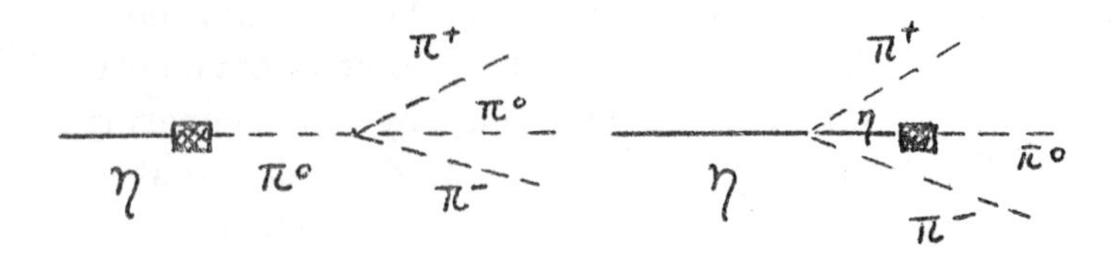

Fig. 3.

for a mass dependence of the η - π^0 transition,[57] or for some symmetry breaking in the Lagrangian (89), then a non-zero result can be obtained. One might however wonder whether such variation could provide the necessary strength for the copious $\eta \to 3\pi$ rate observed. In any case, it won't do any good to the R ratio.

Let us now outline Sutherland's argument. The decay $\eta \to 3\pi$ proceeds via a second order electromagnetic interaction, whose current $J_\mu = J_\mu^{(s)} + J_\mu^{(v)}$ is made of isoscalar and isovector parts. As the $\eta \to 3\pi$ decay occurs with change in G-parity, there is no contribution to the decay from terms like $J_\mu^{(s)} J_\nu^{(s)}$ and $J_\mu^{(v)} J_\nu^{(v)}$. Then the $\eta \to \pi_i\pi_j\pi_k$ amplitude reads:

$$M = 2e^2 < \pi^i\pi^j\pi^k \mid \int d^4y\, D_{\mu\nu}(y)\, T\,[J_\mu^{(0)}(y)\, J_\nu^{(1)}(0)\,] \mid \eta > \tag{90}$$

where $D_{\mu\nu}$ is the propagator of the virtual photon involved in the transition. In the limit with one of the pions four momenta vanishing one obtains

$$M(q_{\pi k} \to 0) \propto e^2 <\pi^i \pi^j | \int d^4y \, D_{\mu\nu}(y)$$

$$\times \, T \{ [{}^5Q^k(y_0), J_\mu^{(5)}(y)] \, J_\nu^{(v)}(0)$$

$$+ J_\mu^{(5)}(y) \, [{}^5Q^k(0), \, J_\nu^{(v)}(0) \,] \} | \eta > . \qquad (91)$$

The equal time commutators involved give according to the current algebra

$$[{}^5Q^k(y_0), \, J_\mu^{(5)}(y)] = 0 \qquad (92)$$

$$[{}^5Q^k(0), \, J_\nu^{(v)}(0)] = i\epsilon_{k31} A_\nu^1(0) \quad .$$

Hence, one is left with an isovector operator sandwiched between η and a two-pion supposedly S-wave which can be in a $I = 0$ or 2 isotopic spin state. This gives zero for (91) and hence one concludes that the $\eta \to 3\pi$ amplitude vanishes whenever any of the three pion momenta tends to zero. This means, by using $s = (q_{\pi^+} + q_{\pi^-})^2$, that $M_{\eta \to 3\pi} = 0$ for $s = m_\eta^2$ and $s = m_\pi^2$. If the matrix element is assumed to be linear in s, this implies the vanishing of the amplitude everywhere.

Itzykson, Jacob and Mahoux[58] take Sutherland's result as an indication that the matrix element is not linear and suggest

$$A_{\eta \to \pi^+ \pi^- \pi^0} = a(s - m_\pi^2) \, (m_\eta^2 - s)$$

$$A_{\eta \to 3\pi^0} = a \{ (s - m_\pi^2)(m_\eta^2 - s) + (t - m_\pi^2) \, (m_\eta^2 - t) \qquad (93)$$

$$+ (u - m_\pi^2)(m_\eta^2 - u) \} \quad .$$

With constant a, one has

$$R = \frac{\eta \to 3\pi^0}{\eta \to \pi^+\pi^-\pi^0} = 1.62.$$

If a is chosen to be an empirical function of s, so as to account for the experimental Dalitz plot, namely $a(s) = 1 + 0.227(s/m_\pi^2)$, one has R = 1.53, still far away from the experimental value.

There have been several works which tried to overcome this unpleasant result. Das, Grynberg and Kikkawa[59] obtain a non-zero result for $\eta \to 3\pi$ by a different treatment of the so-called Schwinger terms. In so doing, they depart from the usual treatment of $K \to 3\pi$ decays, so that the unity of treatment of these similar processes is destroyed. Moreover, the slope they obtain for the spectrum of π^0 is off the experimental value by a factor of 2 and the absolute value obtained for $\eta \to 3\pi^0$ is 120 eV only.

Another group of authors[60-63] obtain a non-zero decay rate by using the ad hoc assumption that the effective electromagnetic interaction mediating $\eta \to 3\pi$ transforms like the third component of an octet of scalar densities.

A positive turn in this confusing situation has been achieved recently by Bardeen, Brown, Lee and Nieh[64] and Dolgov, Vainshtein and Zacharov.[65] These authors start with symmetric off-shell amplitude for

$$\eta\,(p) \to \pi^\alpha(q_1) + \pi^\beta(q_2) + \pi^\gamma(q_3)$$

$$T^{\alpha\beta\gamma}(q_1, q_2, q_3) =$$

$$= (2\pi)^{3/2} (2M_\eta)^{1/2} \left[\prod_{i=1}^{3} (q_i^2 - \mu^2) \int d^4x_i \right.$$

$$\left. \exp(iq_i x_i) \right] \tag{94}$$

$$< 0 | T\{ D^\alpha(x_1)\, D^\beta(x_2)\, D^\gamma(x_3)\, H_{em}(0) \} | \eta >$$

where H_{em} is the same as that used by Sutherland[53] and

$$D^\alpha(x) \equiv \partial^\mu A_\mu^\alpha(x).$$

The non-zero result is obtained by carefully following the variation of pion momenta from zero (where the current algebra statements are being made) to the physical values $q_i^2 = \mu^2$. The variation q_i is performed with the constraint $p = \Sigma q_i$. The amplitude is written as

$$T^{\alpha\beta\gamma}(q_1, q_2, q_3) = \delta_{3\alpha}\, \delta_{\beta\gamma}\, A_1(s_i, Z_i)$$

$$+ \delta_{3\beta}\, \delta_{\gamma\alpha}\, A_2(s_i, Z_i)$$

$$+ \delta_{3\gamma}\, \delta_{\alpha\beta}\, A_3(s_i, Z_i) \tag{95}$$

with $s_i = (p - q_i)^2$, $Z_i = q_i^2$. Assuming linear dependence on s_j and Z_j and using Bose symmetry one has

$$A_1 = C_0 + C_1 (s_2 + s_3) + C_2 s_1 + C_3 (Z_2 + Z_3) \tag{96}$$

with A_2, A_3 given by cyclic permutation of s_i, Z_i. Using in (94) the commutation relations

$$[{}^5Q_\alpha (y_0), D_\beta(y)] = i\delta_{\alpha\beta}\, \sigma(y) \tag{97}$$

$$[{}^5Q_\alpha (y_0), \sigma(y)] = -i D_\alpha (y)$$

and taking the limit $q_1 \to 0$ one finds that $A_1 \to 0$ as $q_1 \to 0$ and $A_2 = A_3 = 0$ as $q_1 \to 0$, Z_2, $Z_3 \to \mu^2$. These conditions lead to amplitudes

$$A_i = C_1 [M_\eta^2 - s_i + (1 - M_\eta^2/\mu^2) Z_i]$$

which on the mass shell are

$$A_i(s_i) = -C_1 M_\eta^2 \left(1 - \frac{2E_i}{M_\eta}\right) \tag{98}$$

where E_i is the energy of the pion i in the η rest system. The slope obtained, $a = -2(M_\eta - 3\mu)M_\eta = -0.49$ agrees very well with the linear fit to the η spectrum giving $a_{exp} = -0.478 \pm 0.038$. The definition of the slope is $M^2\alpha(1 + 2ay)$, where $y = (T - \bar{T}/T$, T being the kinetic energy of the odd pion and $\bar{T}$ its mean value. The authors of Ref. 64 then determine an upper limit for C_1, obtaining $\Gamma_\eta(3\pi^0) \lesssim 160$ eV.

This seems again to be somewhat on the lower side. In addition, the ratio of $3\pi^0/\pi^+\pi^-\pi^0$ is as given by the symmetric isotopic wave function for $I = 1$, and hence quite remote from the reported experimental value.

Before concluding this section, we mention another model which has been suggested originally by Oneda, Kim and Kaplan[66] and reconsidered recently by C. H. Woo.[67] The "nearly vanishing" of the $\eta \to 3\pi$ decay to S-wave pions is taken seriously and it is suggested that the P-wave contribution could be as large as the diminished S-wave part. Woo[67] shows that by judiciously choosing the free parameters of the model one can fit the Dalitz plot and obtain a value for R as low as 1.1. The obvious question still remains, why should $\eta \to 3\pi$ occur at such significant rate if its decay mechanism is due mainly to waves seriously inhibited by angular momentum barriers.

4. Radiative Decays

One of the few successes in the realm of η decays is the calculation of the $\Gamma_\eta(\pi^+\pi^-\gamma)/\Gamma_\eta(\gamma\gamma)$ ratio. This was done[28,48] several years ago by using the vector dominance model of Gell-Mann, Sharp and Wagner. In Ref. 48 one starts from the vertex $\eta\rho\gamma$ and then ρ is allowed to decay into $\pi\pi$ or γ. The $\eta\rho\gamma$ black-box cancels out in taking the $\pi^+\pi^-\gamma/\gamma\gamma$ ratio, and assuming the direct coupling $\rho \to \gamma$ to be given by

$$\frac{em_\rho^2}{g_{\rho\pi\pi}}$$

(implying ρ dominance of the pion form factor) one obtains a ratio of 0.13 for the above ratio. Hence, the three body phase space for a pair of P-wave pions plus a γ-ray achieves the lowering of the $\eta \to \pi^+\pi^-\gamma$ decay below the rate of the higher order $\eta \to 2\gamma$ mode.

As it was explained in the last chapter the current algebra calculations[29, 30] give a result very close to the above one.

One should correct this calculation for other possible, intermediate states. Namely, starting the dispersion analysis with the Yukawa PVV vertex (P-pseudoscalar octet, V-vector nonet), one has for $\eta \to \pi\pi\gamma$ the chain $\eta \to \rho\rho \to \pi\pi\gamma$, while for $\eta \to \gamma\gamma$ one has $(\eta \to \rho\rho \to \gamma\gamma) + (\eta \to \omega\omega \to \gamma\gamma) + (\eta \to \varphi\varphi \to \gamma\gamma) + (\eta \to \omega\varphi \to \gamma\gamma)$. When one includes the additional states for $\eta \to \gamma\gamma$ using SU_3 symmetry + $(\omega - \varphi)$ mixing for relating the couplings, one obtains for $\beta = (\eta \to \pi^+\pi^-\gamma)/(\eta \to \gamma\gamma)$ a value 0.24 according to Faier[68] or 0.14 according to Kim, Oneda and Pati.[69] The difference between the two calculations lies in the way the SU_3 relations are enforced for vector meson $\leftrightarrow \gamma$ couplings. Defining

$$f_{\omega_8\gamma} = \frac{em^2_{\omega_8}}{f_{\omega_8}} \quad \text{and} \quad f_{\rho\gamma} = \frac{em^2_{\rho}}{f_{m_\rho}},$$

Faier uses the procedure of Dashen and Sharp[70] of relating

$$f_{\omega_8} = \sqrt{3}\, f_\rho$$

and then uses the actual physical masses in the couplings of ω and φ. On the other side, Kim, Oneda and Pati use SU_3 in the form

$$f_{\omega_8\gamma} = \frac{1}{\sqrt{3}} f_{\rho\gamma} .$$

In Faier's case, the result turns out to be the same as using only SU_3 without any ω - φ mixing. In the calculation of Kim, Oneda and Pati, the result is only slightly dependent on the mixing angle θ in the range $\theta \simeq 20^0 - 40^0$, and the result comes close to the ρ-dominant model of Brown and Singer[48] ($\beta = 0.13$) as well as to the experimental value $\beta_{exp} = 0.15 \pm 0.035$.

The radiative mode $\eta \to \pi^0\gamma\gamma$ is another of the troublesome spots. Its apparent copious occurrence constitutes somewhat of a theoretical embarrassment. Of the models we have discussed only the A-quantum number classification would predict a rate for $\eta^0 \to \pi^0 \gamma\gamma$ which is comparable to $\eta \to \gamma\gamma$ and $\eta \to 3\pi$. If one estimates the $\eta \to \pi^0\gamma\gamma / \eta \to \pi^+\pi^-\gamma$ ratio by using a ρ-dominance model, one obtains $\eta \to \pi^0\gamma\gamma/\eta \to \pi^+ \pi^- \gamma \simeq \rho^0 \to \pi^0\gamma/\rho^0 \to \pi^+\pi^- < 1\%$. Alles, Baracca and Ramos[71] have refined this by including all possible vector-meson intermediate states and obtain $\eta \to \pi^0\gamma\gamma/\eta \to \gamma\gamma = 1.06 \times 10^{-3}$. This is to be compared with the experimental ratio of approximately 0.6.

When the current algebra and PCAC are applied to this decay, one finds that the amplitude should vanish for the point

$$E_{\gamma 1} = E_{\gamma 2} = \frac{m_\eta}{2} .$$ [72]

With a simple matrix element $CF_{\mu\nu} F^{\mu\nu}$, this would imply the vanishing of the amplitude everywhere. Hence, we are again with a case which looks vanishingly small theoretically and very potent experimentally.

It was recently suggested[45] to look for a solution by taking into account quadrilinear meson-meson interactions, as the trilinear ones, which helped in accounting for the other η radiative decays give a very small rate in this case. If a term generalizing in a SU_3 sense the well-known $\lambda(\vec{\pi}\cdot\vec{\pi})^2$ interaction is introduced into the interaction Lagrangian, it contains also a part

$$L_{int} = \frac{g}{2\mu^2} \eta \vec{\pi} \cdot \vec{\rho}_{\mu\nu} \chi^{\mu\nu} \tag{99}$$

where χ is the appropriate combination of ω and φ which has octet SU_3 properties. Introducing into this Lagrangian the coupling to the electromagnetic field, one obtains the $\eta \to \pi^+\pi^-\pi^0\gamma$ and $\eta \to \pi^0\gamma\gamma$ decays through the chains

(1) $\eta \to (\pi + \rho + \chi) \to \pi^+ \pi\pi^+ \gamma$

(2) $\eta \to (\pi^+ \rho^+ \chi) \to \pi^+ \gamma^+ \gamma.$

A ratio between these two modes can be obtained (where the g/μ^2 constants drop out) by using this model:

$$\Gamma(\eta \to \pi^+ \pi^- \pi^0 \gamma) / \Gamma(\eta \to \pi^0 \gamma\gamma) = 0.23\% \tag{100}$$

Experimentally one has $\Gamma(\eta \to \pi^+ \pi^- \pi^0 \gamma / \Gamma(\eta \to \pi^+ \pi^- \pi^0)$ $< 0.9\%$, and given the approximate experimental equality of the $\eta \to \pi^0 \gamma\gamma$ and $\eta \to \pi^+ \pi^- \pi^0$ modes, this is consistent with Singer's model. One should add that this ratio can also be obtained by the current algebra technique which leads to Eq. (84), by taking $A = (\eta^0)$, $B = (\pi^0 \gamma)$. Due to the remoteness of the P-wave pion pair from the ρ-mass region, the equivalence of the two methods holds very well in this case.

Concerning the rate for $\eta \to \pi^0 \gamma\gamma$, it turns out that one has to take $g^2/4\pi \simeq 30$ in order to accommodate this mode within an η life time of approximately 2 KeV. This is somewhat on the higher side, though not unreasonable.

5. Leptonic Decays

The decays we call "leptonic" are those decays in which one or both gamma rays from a radiative η-decay are replaced by $e^+ e^-$ or $\mu^+ \mu^-$ Dalitz pairs. They are obviously expected to be rare. Their interest lies in the possibility of learning about the structure of the vertex functions from which these decays derive - the structure of the vertex functions being determined by the strong interactions. The decays $\eta \to \gamma^0 e^+ e^-$, $\gamma^0 \mu^+ \mu^-$, $\mu^+ \mu^- e^+ e^-$, $e^+ e^- e^+ e$, $\mu^+ \mu^- \mu^+ \mu^-$, $\mu^+ \mu^-$ and $e^+ e^-$ have been discussed recently by Celeghini and Gatto,[73] Geffen and Bing-lin Young,[74] Jarlskog and Pilkuhn,[75] Callan and Treiman[76] and most extensively by Bing-lin Young.[77]

Let us denote by $\Gamma_\eta(k_1^2, k_2^2)$ the vertex function for the process $\eta \to \gamma\gamma$. Then the amplitudes for $\eta \to 2\gamma$, $\eta \to \gamma 1^+ 1^-$, $\eta \to 1^+ 1^-$, $\eta \to 1^+{}_1 1^-{}_2 1^+{}_3 1^-{}_4$, can be written respectively as:

$$A_{\eta\to 2\gamma} = \epsilon_1^{\mu}\,\epsilon_2^{\nu}\,\epsilon_{\mu\nu\sigma\tau}\frac{k_1^{\sigma}\,k_2^{\tau}}{m_{\pi}}\,\Gamma_{\eta}(0,0) \tag{100b}$$

$$A_{\eta\to\gamma 1^+1^-} = -ie\,\bar{u}(p_2)\,\gamma^{\mu}v(p_1)\,\frac{1}{(q-k)^2}$$

$$\times\,\epsilon^{\nu}\,\epsilon_{\mu\nu\sigma\tau}\,\frac{k^{\sigma}q^{\tau}}{m_{\pi}}$$

$$\times\,\Gamma_{\eta}(0,\ (q-k)^2) \tag{101}$$

$$A_{\eta\to 1^+1^-} = -\frac{e^2}{(2\pi)^4}\,\bar{u}(p_2)\,\gamma^{\nu}\gamma^{\mu}v(p_1)\,\epsilon_{\mu\nu\sigma\tau}$$

$$\times\,\frac{1}{m_{\pi}}\int d^4p\,\frac{(m_1+\not{p})(p_1+p)^{\sigma}(p_2-p)^{\tau}}{(p^2-m_1^2+i\epsilon)[(p_1+p)^2+i\epsilon][(p_2-p)^2+i\epsilon]}$$

$$\times\,\Gamma_{\eta}[(p_1+p)^2,\ (p_2-p)^2]\,. \tag{102}$$

The form factor $\Gamma_\eta(k_1^2,\ k_2^2)$ is defined by

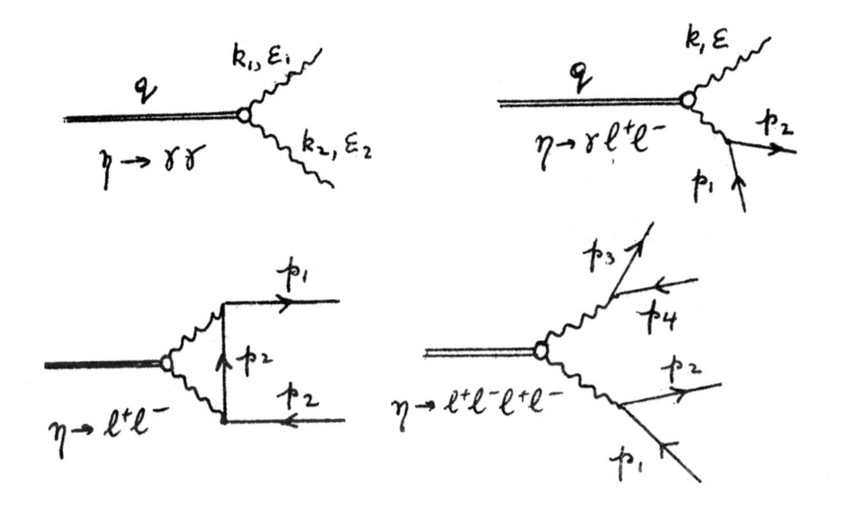

Fig. 4.

$$\Gamma_{\mu\nu}(k_1, k_2) = \epsilon_{\mu\nu\sigma\tau} \frac{k_1^{\sigma} k_2^{\tau}}{m_{\pi}} \Gamma_{\eta}(k_1^2, k_2^2) = \tag{104}$$

$$= i \int d^4 x e^{i(k_1 - k_2) \cdot x/2} < 0 | T\{ J_{\mu}(x/2), J_{\nu}(-x/2)\} | q >$$

where the notation can be read from Fig. 4.

$$A_{\eta \to 1^+ 1^- 1^+ 1^-} = \frac{-e^2 [\bar{u}(p_2) \gamma^{\mu} v(p_1)][\bar{u}(p_3) \gamma^{\nu} v(p_4)]}{(p_1 + p_2)^2 [q - (p_1 + p_2)]^2} \times \tag{103}$$

$$\epsilon_{\mu\nu\sigma\tau} \frac{(p_1 + p_2)^{\sigma} q^{\tau}}{m_{\pi}} \Gamma_{\eta} ((p_1 + p_2)^2, [q - (p_1 + p_2)]^2) .$$

The ratios of the various leptonic decays to $\Gamma_{\eta \to 2\gamma}$ depend of course on $\Gamma_\eta(k_1^2, k_2^2)$. The variation of this function according to various models has been analyzed in the references mentioned above.

For the decay $\eta \to \mu^+ \mu^-$, $e^+ e^-$ a lower limit can be obtained, as the imaginary part of the amplitude can be obtained by setting the two photon lines on the mass shell. The lower limit thus obtained is essentially an expression of the use of unitarity. Some typical numbers[75] for the ratios expected are

$$\Gamma_\eta(e^+ e^- \gamma)/\Gamma_\eta(2\gamma) \simeq 0.016;$$

$$\Gamma_\eta(\mu^+ \mu^- \gamma)/\Gamma_\eta(2\gamma) \simeq 5 \times 10^{-4};$$

$$\Gamma_\eta(e^+ e^- e^+ e^-)/\Gamma_\eta(2\gamma) \simeq 6.6 \times 10^{-5};$$

$$\Gamma_\eta(e^+ e^- \mu^+ \mu^-)/\Gamma_\eta(2\gamma) \simeq 4 \times 10^{-6};$$

$$\Gamma_\eta(\mu^+ \mu^-)/\Gamma_\eta(\gamma\gamma) \geq 0.8 \times 10^{-5};$$

$$\Gamma_\eta(e^+ e^-)/\Gamma_\eta(\gamma\gamma) \geq 4.5 \times 10^{-9}.$$

These numbers change of course when using various models[77] for estimating $\Gamma_\eta(k_1^2, k_2^2)$.

The decay $\eta \to \pi^+\pi^- 1^+ 1^-$ can be related in a similar manner to $\eta \to \pi^+ \pi^- \gamma$ and one obtains

$$\Gamma_\eta(\pi^+ \pi^- e^+ e^-)/\Gamma_\eta(\pi^+ \pi^- \gamma) = 0.0066.$$

As the decay $\eta^0 \to \pi^0 \gamma\gamma$ seems to occur at a remarkable rate one can study in a similar manner the form factor $F_{\eta\pi}(k_1^2, k_2^2)$ in the processes $\eta \to \pi^0 \gamma\gamma$, $\eta \to \pi^0 1^+ 1^- \gamma$, $\eta \to \pi^0 1^+ 1^-$, $\eta \to \pi^0 1^+ 1^- 1^+ 1^-$. Smith[84] has analyzed recently the decays $\eta \to \pi^0 e^+ e^-$, $\eta \to \pi^0 \mu^+ \mu^-$. The lower limits imposed by unitarity in this case turn out to be:

$$\Gamma_\eta(\pi^0 \mu^+ \mu^-)/\Gamma_\eta(\pi^0 2\gamma) \geq 1.0 \times 10^{-5};$$

$$\Gamma_\eta(\pi^0 e^+ e^-)/\Gamma_\eta(\pi^0 2\gamma) \geq 1.3 \times 10^{-8}.$$

The estimate of these processes is of special interest, as the proposed[40] C-violation in electromagnetic interaction would allow the decay $\eta \to \pi^0 1^+ 1^-$ to order α^2.

6. η Life-Time

It has been undertaken recently[78] to measure the η life time by using the Primakoff effect. In particular, the strength of the $\eta \to 2\gamma$ transition is estimated from forward Coulomb photoproduction of η-mesons. Preliminary results indicate a total η-width of a few KeV. This has caused some worries among theorists (see e.g. Ref. 72), as most of the existing calculations predicted a life time of the order of a few hundred electron-volts. Especially, one is wondering about a certain SU_3 calculation which fails to predict this figure.

Let us review shortly the existing estimates. The $\eta \to 2\gamma$ decay mode can be related to the $\pi^0 \to 2\gamma$ by using unitary symmetry. Cabibbo and Gatto[79] have derived the relation for the amplitudes

$$A(\eta \to 2\gamma) = \frac{1}{\sqrt{3}} A(\pi^0 \to 2\gamma) \tag{105}$$

assuming η, π^0 belong to an octet. This relation can be easily obtained, for instance, by forming the two U-spin eigenstates from π^0 and η^0 and noting that the transition between the U-spin one eigenstate and a 2γ state is zero. So, a relation between $\langle \eta/2\gamma \rangle$ and $\langle \pi^0/2\gamma \rangle$ is obtained. The consequent relation between the widths is

$$\Gamma(\eta \to 2\gamma) = \frac{1}{3} \left(\frac{m_\eta}{m_{\pi^0}}\right)^3 \Gamma(\pi^0 \to 2\gamma). \tag{106}$$

With a decay width $\Gamma_{\pi^0 \to 2\gamma} = 7.4 \pm 1.5$ eV[27] one expects from (106) $\Gamma_{\eta \to 2\gamma} = 165 \pm 33$ eV. Okubo and Sakita have used the SU_3 relations obtained in Ref. (79):

$$\langle K^+/JJ/K^+ \rangle = \langle \pi^+/JJ/\pi^+ \rangle \tag{107}$$

$$\langle K^0/JJ/K^0 \rangle = \langle \pi^0/JJ/\pi^0 \rangle - \sqrt{3} \langle \eta/JJ/\pi^0 \rangle$$

to obtain

$$\gamma = \frac{1}{\sqrt{3}} \left[(m_{K^0}^2 - m_{K^+}^2) + (m_{\pi^+}^2 - m_{\pi^0}^2) \right]$$

$$= -(54 \text{ MeV})^2 \qquad (108)$$

where γ is the strength of the effective transition $\eta\pi$. Using the pole model to estimate $\eta \to \pi \to 3\pi$ and a strength $\lambda = -0.18 \pm 0.05$ for the $4\pi\lambda(\vec{\pi} \cdot \vec{\pi})^2$ interaction, they obtain

$$\Gamma_\eta (\pi^+ \pi^- \pi^0) = 147^{+90}_{-70} \text{ eV}.$$

Unfortunately, the adding of the additional diagram required by SU_3, namely $\eta \to \eta\pi\pi \to \pi\pi\pi$ gives a vanishing result[56] for $\eta \to 3\pi$ in the pole model and the significance of the result obtained by Okubo and Sakita is henceforth questionable.

Barrett and Barton have performed a dynamical calculation of $\eta \to 3\pi$, by using intermediate baryon states to calculate the mass dependence of the η-π^0 transition.[57] Their result is

$$\Gamma_{\eta \to \pi^+\pi^-\pi^0} \simeq 140\text{-}280 \text{ eV},$$

not very different from the original Okubo and Sakita calculation.

In the σ-model of Brown and Singer, the width for $\eta \to 3\pi$ depends on the mass and width of σ, as well as on the effective coupling $(\eta\,\sigma\,\pi)$. For $M_\sigma = 400$ MeV, $\Gamma_\sigma = 100$ MeV, one obtains[49]

$$\Gamma_{\eta \to \pi^+\pi^-\pi^0} = 265 \; G^2 \text{ eV},$$

where G^2 is the ratio between the $(\eta\ \sigma\ \pi)$ coupling and α^2 times the $(\sigma\ \pi\pi)$ coupling. No detailed estimate of G has been attempted. Nevertheless, for the reasonable value $G^2 \simeq 1$, one has again

$$\Gamma_{\eta\to\pi^+\pi^-\pi^0} \simeq 250 \text{ eV}.$$

All the calculations just described, would end in a total eta life time of the order of 0.5 KeV.

A recent calculation by Berends and Singer[81] predicts an η-life-time some 4-5 times shorter. In this work, the $\eta\to\pi^+\pi^-\gamma$ transition is calculated assuming it proceeds through the $\eta\ \rho\ \gamma$ vertex. By using SU_3 relations for the $\langle V\ |J|\ P\rangle$ transition as well as $\omega - \varphi$ mixing with an angle $\sin\theta = 1/\sqrt{3}$, one obtains the relation

$$f_{\rho\eta\gamma} = \frac{1}{\sqrt{3}}\ f_{\omega\pi\gamma}\ . \tag{109}$$

Hence, the η life time is related in this work to the $\omega\to\pi\gamma$ transition. For a measured value of $\Gamma_\omega(\pi^0\gamma) = 1.3$ MeV, one obtains $\Gamma_\eta(\pi^+\pi^-\gamma) = 120$ eV which leads to

$$\Gamma_\eta \text{ (total)} = 2.2 \text{ KeV} \tag{110}$$

$$\tau_\eta = 3.0\times 10^{-19} \text{ sec}\ .$$

This result seems to be in good agreement with the existing experimental information. One should mention, however, that the above calculation is sensitive to η-X^0 mixing, which would affect the relation (109), as well as (105).

In the current-algebra calculation of $\eta \to 3\pi$ by Bardeen, Brown, Lee and Nieh[64] there is also an evaluation of the absolute rate. This is obtained in the limit $q_1, q_2 \to 0$ (keeping $\sum_i q_i = p$). The expression for the coefficient of the $\eta \to 3\pi$ amplitude is

$$C = (2\pi)^{3/2} (2m_\eta)^{1/2} \frac{m_\pi^6}{m_\eta^2} \int d^4x \, e^{ip \cdot x}$$

$$< 0 \,|T\{ D^3 (x) \; H_{em} (0)\} \; |\eta > \quad (111)$$

This expression may be related to $\pi^+\pi^0$, $K^+ K^0$ mass differences by SU_3 (eq. 108). By using the pion pole approximation an upper limit for C is obtained

$$\Gamma(\eta \to 3\pi^0) \leq 1.6 \times 10^2 \text{ eV},$$

which seems again on the lower side.

Before ending we remark that Quark model calculations have also been performed to estimate the η life time. Pietschmann and Thirring[82] use essentially the idea of Berends and Singer[81] of calculating the $\eta \to \pi\pi\gamma$ decay mode through the $\eta \to (\rho)\, \gamma$ transition. The $\eta\rho\gamma$ vertex is related to the $\omega\pi\gamma$ by using the quark model, the relation being the same as (109). As well known the quark model succeeds in accounting for the observed $\omega \to \pi\gamma$ width and hence the $\eta \to \pi^+\pi^-\gamma$ width one obtains in the quark model is 120 eV, the same as in the Berends-Singer model. On the other hand, by using the quark model for calculating the $\eta \to 3\pi$ and $\eta \to \pi^0 + 2\gamma$ decays, Möbius and Pietschmann[83] found rates of the order of 1 eV only.

REFERENCES

1. Proceedings of the Liperi Summer School in Theoretical Physics, Aug. 1966, part A (University of Helsinki).
2. Y. Nambu, Phys. Rev. Letters 4, 380 (1960); M. Gell-Mann and M. Lévy, Nuovo Cimento 16, 705 (1960).
3. J. Bernstein, S. Fubini, M. Gell-Mann and W. Thirring, Nuovo Cimento 17, 757 (1960).
4. M. L. Goldberger and S. B. Treiman, Phys. Rev. 109, 193 (1958).
5. S. Adler, Phys. Rev. 137, B1022 (1965); ibidem, 137, B1638 (1965).
6. G. F. Chew and S. Mandelstam, Phys. Rev. 119, 467 (1960).
7. N. H. Fuchs, Phys. Rev. 149, 1145 (1965).
8. See for details: G. F. Chew, "S-Matrix Theory of Strong Interactions" (W. A. Benjamin, Inc., N. Y. 1961), p. 95.
9. L. M. Brown and P. Singer, Phys. Rev. Letters 8, 460 (1962); 16, 424 (1966).
10. S. L. Adler, Phys. Rev. 140, B736 (1965).
11. K. Kawarabayashi, W. D. McGlinn and W. N. Wada, Phys. Rev. Letters 15, 897 (1965).
12. I. J. Muzinich and S. Nussinov, Phys. Letters 19, 248 (1965).
13. S. L. Adler, Phys. Rev. Letters 14, 1051 (1965); also reference 10.

14. W. I. Weisberger, Phys. Rev. Letters 14, 1047 (1965).
15. For full details see References 10 and 11.
16. M. Gell-Mann, Physics 1, 63 (1964).
17. C. K. Pandit and J. Schechter, Phys. Letters 19, 56 (1965).
18. V. S. Mathur and L. K. Pandit, Phys. Rev. 143, 1216 (1966).
19. A. N. Kamal, Phys. Rev. 150, 1392 (1966).
20. S. Weinberg, Phys. Rev. Letters 17, 616 (1966).
21. The formula for scattering of pions on a heavy target has been derived independently by several other authors. See Footnote (5) of Ref. (20).
22. N. Fuchs, Phys. Rev. 155, 1785 (1967).
23. N. Khuri, Phys. Rev. 153, 1477 (1967).
24. The B and C amplitudes are obtained from (65) by exchanging s-t and s-u, respectively.
25. J. Sucher and C. -H. Woo, Phys. Rev. Letters 18, 723 (1967).
26. K. Kawarabayashi and Mahiko Suzuki, Phys. Rev. Letters 16, 255 (1966).
27. A. H. Rosenfeld, A. Barbaro-Galtieri, W. J. Podolski, L. R. Price, P. Soding, C. G. Wohl, M. Roos and W. J. Willis, Rev. Mod. Phys. 39, 1 (1967).
28. M. Gell-Mann, D. Sharp and W. G. Wagner, Phys. Rev. Letters 8, 261 (1962).
29. M. Ademollo and R. Gatto, Nuovo Cimento 44, 282 (1966) .
30. J. Pasupathy and R. E. Marshak, Phys. Rev. Letters 17, 888 (1966).
31. P. D. Conway, Phys. Letters 24 B, 59 (1967).
32. P. Singer, Phys. Rev. 128, 2789 (1962).
33. J. J. Sakurai, Ann. of Phys. 11, 1 (1960).

34. J. J. Sakurai, Ann. of Phys., to be published.
35. M. Ademollo, Nuovo Cimento 46, 156 (1966).
36. F. S. Crawford and L. R. Price, Phys. Rev. Letters 16, 333 (1966).
37. C. R. Kalbfleisch, O. I. Dahl and A. Rittenberg, Phys. Rev. Letters 13, 349 A (1964).
38. H. Rubinstein and G. Veneziano, Phys. Rev. Letters 18, 411 (1967).
39. A. Pevsner et al, Phys. Rev. Letters 7, 421 (1961).
40. J. Bernstein, G. Feinberg and T. D. Lee, Phys. Rev. 139, B1650 (1965).
41. F. A. Berends and P. Singer, Nuovo Cimento 46, 90 (1966).
42. L. R. Price and F. S. Crawford, Jr., Phys. Rev. Letters 18, 1207 (1967).
43. M. Feldman, W. Frati, R. Gleeson, J. Halpern, M. Nussbaum and S. Richert, Phys. Rev. Letters 18, 868 (1967).
44. J. B. Bronzan and F. Low, Phys. Rev. Letters 12, 522 (1964).
45. P. Singer, Phys. Rev. 154, 1592 (1967).
46. See, e.g. G. Källen: "Elementary Particle Physics" p. 430.
47. M. Veltman and J. Yellin, Phys. Rev. 154, 1469 (1967).
48. L. M. Brown and P. Singer, Phys. Rev. Letters 8, 460 (1962); 16, 424 (1966).
49. L. M. Brown and P. Singer, Phys. Rev. 133, B812 (1964).
50. A. Donnachie, R. M. Heinz and C. Lovelace, Phys. Letters 22, 332 (1966).
51. K. C. Wali, Phys. Rev. Letters 9, 120 (1962).

52. B. Barrett, M. Jacob, M. Nauenberg and T. N. Truong, Phys. Rev. 141, 1342 (1966).
53. D. G. Sutherland, Phys. Letters 23, 384 (1966).
54. M. A. B. Bég, Phys. Rev. Letters 9, 67 (1962).
55. C. Kacser, Phys. Rev. 130, 355 (1963).
56. S. Hori, S. Oneda, S. Chiba, and A. Wakasa, Phys. Letters 5, 339 (1963).
57. B. Barret and G. Barton, Phys. Rev. 133, B466 (1964).
58. C. Itzykson, M. Jacob and G. Mahoux, Ann. Phys. (to be published).
59. T. Das, M. Grynberg and K. Kikkawa, Phys. Rev. 156, 1568 (1967).
60. S. K. Bose and A. M. Zimerman, Nuovo Cimento 43, 1165 (1966).
61. R. Ramachandran, Nuovo Cimento 47 A, 669 (1967).
62. Y. T. Chiu, J. Schechter and Y. Ueda, Phys. Rev. 161, 1612 (1967).
63. R. H. Graham, L. O'Raifeartaigh and S. Pakvasa, Nuovo Cim. 48A, 830 (1967).
64. W. A. Bardeen, L. S. Brown, B. W. Lee and N. T. Nieh, Phys. Rev. Letters 18, 1170 (1967).
65. A. D. Dolgov, A. I. Vainshtein and V. I. Zacharov, Phys. Letters 24 B, 425 (1967).
66. S. Oneda, Y. S. Kim and L. M. Kaplan, Nuovo Cimento 34, 655 (1964).
67. C. H. Woo, Phys. Rev. 156, 1719 (1967).
68. H. Faier, Nuovo Cimento 41A, 127 (1966); 43A, 230 (1966).
69. Y. S. Kim, S. Oneda and J. C. Pati, Phys. Rev. 135, B1076 (1964).
70. R. F. Dashen and D. H. Sharp, Phys. Rev. 133, B1585 (1964).

71. W. Alles, A. Baracco and A. T. Ramos, Nuovo Cimento 45A, 272 (1966).
72. D. G. Sutherland, Nucl. Phys. 2B, 433 (1967).
73. E. Celeghini and R. Gatto, Nuovo Cimento 28, 1497 (1963).
74. D. A. Geffen and Bing-lin Young, Phys. Rev. Letters 15, 316 (1965).
75. C. Jarlskog and H. Pilkuhn, Nuclear Physics 1B, 264 (1967).
76. C. G. Callan, Jr. and S. B. Treiman, Phys. Rev. Letters 18, 1083 (1967); 19, 57 (1967).
77. Bing-lin Young, Phys. Rev. 161, 1620 (1967).
78. C. Bemporad, P. L. Braccini, L. Foà, L. Lübelsmeyer, D. Schmitz, Phys. Letters 25B, 380 (1967).
79. N. Cabibbo and R. Gatto, Nuovo Cimento 21, 872 (1961).
80. S. Okubo and B. Sakita, Phys. Rev. Letters 11, 50 (1963).
81. F. A. Berends and P. Singer, Phys. Letters 19, 249; 616 (1965).
82. H. Pietschmann and W. Thirring, Phys. Letters 21, 713 (1966).
83. P. Möbius and H. Pietschmann, Phys. Letters 22, 684 (1966).
84. J. Smith, Phys Rev. 166, 1629 (1968).